Computational Music Science

About this series - The CMS series covers all topics dealing with essential usage of mathematics for the formal conceptualization, modeling, theory, computation, and technology in music. The series publishes peer-reviewed only works.

Comprehensiveness - The series comprises symbolic, physical, and psychological reality, including areas such as mathematical music theory, musical acoustics, performance theory, sound engineering, music information retrieval, AI in music, programming, soft- and hardware for musical analysis, composition, performance, and gesture. The CMS series also includes mathematically oriented or computational aspects of music semiotics, philosophy, and psychology.

Quality - All volumes in the CMS series are published according to rigorous peer review, based on the editors' preview and selection and adequate refereeing by independent experts.

Collaboration - The editors of this series act in strong collaboration with the Society for Mathematics and Computation in Music and other professional societies and institutions.

Should an author wish to submit a manuscript, please note that this can be done by directly contacting the series Editorial Board, which is in charge of the peer-review process.

Manuscripts submitted for publication within the series will undergo a careful and deep review process, which might take some time to be completed.

THE SERIES IS INDEXED IN SCOPUS

Guerino Mazzola · Ryan Dornfeld · Ruochen Xi ·
Ran Xu · Junze Gong · Yuwen Sun · Qiu Qiu ·
Micah Winter · Jasmin Bryan-Moreira

Basic Gestures for Music Theory, Performance, and Pedagogy

Springer

Guerino Mazzola
School of Music
University of Minnesota
Minneapolis, MN, USA

Ruochen Xi
School of Music
University of Minnesota
Minneapolis, MN, USA

Junze Gong
School of Music
University of Minnesota
Minneapolis, MN, USA

Qiu Qiu
School of Music
University of Minnesota
Minneapolis, MN, USA

Jasmin Bryan-Moreira
School of Music
University of Minnesota
Minneapolis, MN, USA

Ryan Dornfeld
School of Music
University of Minnesota
Minneapolis, MN, USA

Ran Xu
School of Music
University of Minnesota
Minneapolis, MN, USA

Yuwen Sun
School of Music
University of Minnesota
Minneapolis, MN, USA

Micah Winter
School of Music
University of Minnesota
Minneapolis, MN, USA

ISSN 1868-0305 ISSN 1868-0313 (electronic)
Computational Music Science
ISBN 978-3-032-10680-3 ISBN 978-3-032-10681-0 (eBook)
https://doi.org/10.1007/978-3-032-10681-0

This Springer imprint is published by the registered company Springer Nature Switzerland AG
The registered company address is: Gewerbestrasse 11, 6330 Cham, Switzerland

Dedicated to Casilda Mazzola

Photo and © by Ida Mazzola

Preface

The idea to write this book was first proposed by Jasmin with some of her colleagues at the end of Mazzola's courses on gesture and performance theory. The philosophy of such a book is an educational approach to fundamental topics defining music. Music is viewed as a symbolic entity, typically presented by a score, together with its performative transformation to physical reality of sounds and their creative mechanical or electronic instruments.

This book first develops the music theory of symbolic entities and correlations and then introduces the modes of describing and understanding performance. We propose these approaches in the new language of musical gestures, a methodology that strongly respects the first approach to music for young students.

A third part of the book sketches the challenging pedagogical dimension of teaching music in this elementary spirit.

The co-authors are all my students and will guarantee that the language and methods as displayed here can be digested without higher scientific perspectives. I am in particular grateful for Jasmin's, Mica's, and Ryan's intense interaction along with the book's creation. Thanks from their teacher's, Guerino Mazzola's, part.

We are pleased to acknowledge the support from Springer's science editor Francesca Bonadei for writing such a demanding treatise.

Minneapolis, September 2025

Fig. -1.1: The book's authos (from left top) Jasmin Bryan-Moreira, Junze Gong, Qiu Qiu, Ran Xu, Micah Winter, Yuwen Sun, Ryan Dornfeld, Ruochen Xi, Guerino Mazzola.

Contents

Part V Pedagogy

Part I

What Are Gestures

1

Life Gestures

Summary. This initial chapter introduces the book's general topic, namely gestures in their general conceptualization, which can be found in well-known sources such as *Wikpedia* on the internet. This chapter prepares the readers to the specific musical gesture concept, which is introduced in Chapter 2, and which will be used for the development of a gesturally shaped music and performance theory.

– Σ –

1.1 Common Gestural Concepts

In *Wikpedia* [118], gestures are defined with these words:

> A gesture is a form of nonverbal communication or non-vocal communication in which visible bodily actions communicate particular messages, either in place of, or in conjunction with, speech. Gestures include movement of the hands, face, or other parts of the body. Gestures differ from physical non verbal communication that does not communicate specific messages, such as purely expressive displays, proxemics, or displays of joint attention. Gestures allow individuals to communicate a variety of feelings and thoughts, from contempt and hostility to approval and affection, often together with body language in addition to words when they speak. Gesticulation and speech work independently of each other, but join to provide emphasis and meaning.

In other words, gestures are a form of physical non-verbal communication that can be used to indicate specific information or responses to an ongoing conversation or event. Gestures may be a part of conversation or independent of speech. However, gestures are not necessarily specific signs (elements of semiotic system).

For humans, gestures are a physical form of nonverbal communication, the gesturalizing persons perform movements with their body, typically their

G. Mazzola et al., *Basic Gestures for Music Theory, Performance, and Pedagogy*, Computational Music Science, https://doi.org/10.1007/978-3-032-10681-0_1

hands, arms, and head. Gestures may be accompanied by speech, but gestures often occur without words, see Figure 1.1 for an image of a wordless gesture. One builds the shape of a heart for a short time.

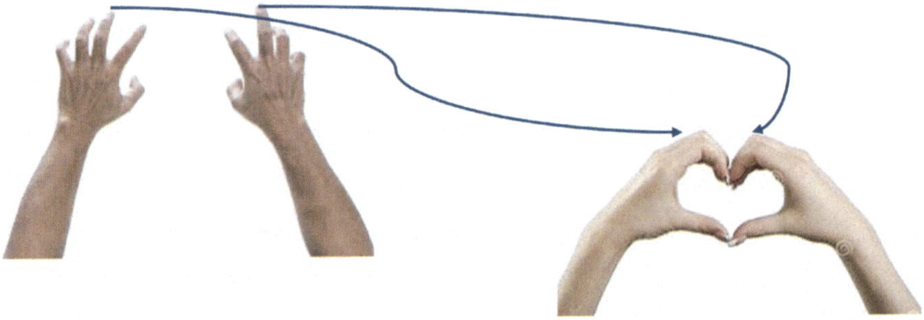

Fig. 1.1: A screenshot of a movement of a common gesture without words.

This short movement expresses a feeling or communication of love. Many gestures bear a meaning, which is typically given by a social convention. The French philosophy of gestures calls such gestures "tamed", in contrast of "untamed" or "wild" gestures, which generate a new content, when performed. For example, a military hinting gesture could designate a new target for an armed attack.

A gesture's movement is embodied by the gesturalizing person, i.e., the movement is realized by the involved and defining limbs. This means that a gesture is a movement in time of physical objects (hands, arms, etc.). It consists of three components:

1. the body's parts,
2. the movement or curve in time, and
3. the embodiment of the time curve through the body.

The body here is often composed of a number of parts that move (hand, arm, fingers, etc.) This means that the curves of a gesture are assigned to specific body parts. A gesture is *articulated* by body parts, which are each given the same number of curves in time.

This is a first description of gestures that the human body performs in the real physical space. In the next chapter, we shall take this template as a model of gestures in the context of music. But we should recognize from recent brain research that the neural activity of collaborating humans, mum and her child for example, is synchronized by their body's gestural collaboration.

2

Musical Gestures

Summary. In this chapter, we develop a concept of gestures as they are used in music.

– Σ –

2.1 The Three Components of a Musical Gesture

Fig. 2.1: A musical score

2.1.1 The Musical Space of a Gesture and its Curves

For the above case of common gestures in real life, we have exhibited the physical space (three space dimensions plus one time dimension), where such

G. Mazzola et al., *Basic Gestures for Music Theory, Performance, and Pedagogy*, Computational Music Science, https://doi.org/10.1007/978-3-032-10681-0_2

gestures move with the body's parts. In music, this space is still possible, such in the case of the gesture of a pianist's fingers on the keyboard, which move in the physical space in time. But often, musical gestures live in spaces that describe the position and movement of musical objects without a defined physical position.

Musical spaces are the domains where musical objects live. We are not going to describe such spaces in mathematical terms, it will be sufficient to specify the properties, which are invoked and specified to address usual musical objects. The most common musical space is what we see, when looking at a musical score, see Figure 2.1.

Such a score shows musical objects, the *notes*, which are round black or white/black spots distributed over staffs. A staff has five horizontal lines, which specify the higher or lower position of a note (on the piano, higher/lower positions are played to the right/left of the keyboard). The musical time goes from left to right on a score. Notes with the same time position are played together as *chords*. The space of a score is not a physical space, it consists of symbols that indicate, how to perform music in physical reality, using different special performance signs, such as tempo (metronome speed information), and dynamics (pianissimo *pp*, mezzoforte *mf*, forte *f*, fortissimo *ff*, etc.).

In a musical space, such as our score, one may consider curves that connect different notes, as shown in Figure 2.2.

Fig. 2.2: A curve in a musical space

Such curves needn't follow musical time, they can connect whatever points. Curves are the new concepts, which generalize movements in physical space. A curve c specifies a point $c(t)$ in space for every number t between 0 and 1, in which 0 is the beginning point and 1 is the endpoint of the curve, in other words, a curve is a continuous map. If X is a musical space, the collection of its curves is denoted by $X^{\rightarrow}$. This is the first component of a musical gesture.

2.1.2 The Skeleton of a Gesture

To describe a gesture, we have to specify the curves' organization. To this end, we are constructing a system of abstract arrows, which we call the gesture's *skeleton*. Figure 2.3 shows a skeleton S. It consists of arrows that may be connected to each other, and may appear in multiple or parallel organization between two inital and final points. Figure 2.3 shows a skeleton of arrows between points A, B, and C.

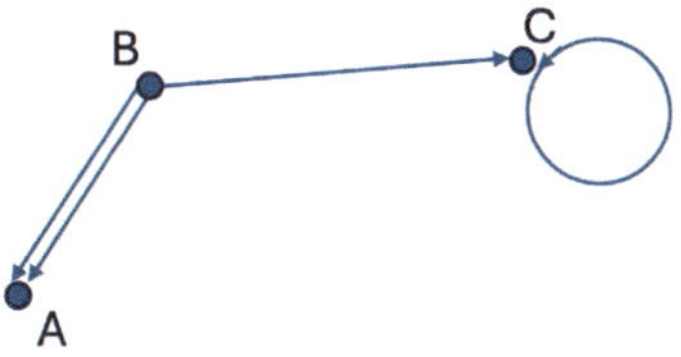

Fig. 2.3: A skeleton with three initial and final points A, B, and C. A loop in C is also shown, and two parallel arrows from B to A.

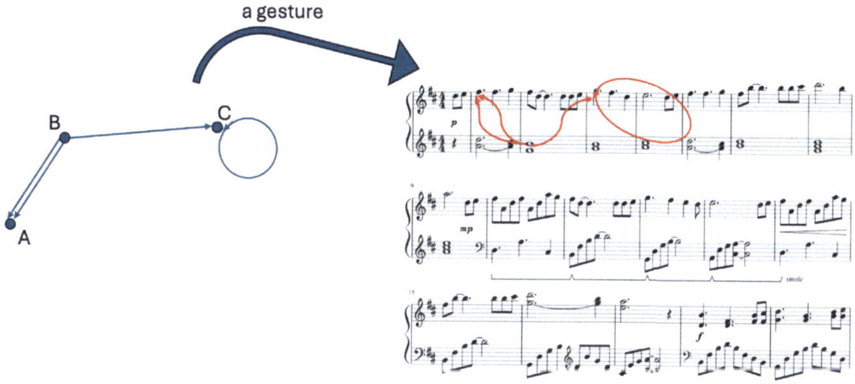

Fig. 2.4: A gesture from the skeleton to the left to the curve space to the right, defined by four curves in red.

2.1.3 Connecting a Skeleton S with the Curves in $X^{\rightarrow}$

A gesture consists of three components:

1. a skeleton S,
2. a curve space $X^{\rightarrow}$, and
3. a rule c that maps every arrow a of S to a curve $c(a)$ in $X^{\rightarrow}$, with the sole condition that if two arrows have a common point, then their curves also have common image points.

See Figure 2.4 for an example. A gesture needn't have any profound meaning or further implication, it is simply a description of the connection between points in music. This example is intended to demonstrate the correlation of a gesture skeleton to a musical space, not to imply any further significance.

2.1.4 Hypergestures

A revolutionary consequence of our definition of a gesture is the fact that we may now consider "gestures of gestures". Let us make this precise. Denote by $S@X^{\rightarrow}$ the space (collection) of all gestures with skeleton S and curves in space X. We may consider curves in this space, intuitively: such a curve would give us a gesture for every curve argument, similar to the curves in $X^{\rightarrow}$. We may then look at gestures in a $T@(S@X^{\rightarrow})$ for a skeleton T, defining the space $T@(S@X^{\rightarrow})$ of *hypergestures*, i.e., of gestures with skeleton T over gestures with skeleton S and curves in space X. See Chapter 3 for first examples.

It is a substantial result in gesture theory that $T@(S@X^{\rightarrow})$ and $S@(T@X^{\rightarrow})$ are essentially the same:

Theorem 1. *(Escher Theorem) For any two skeleta S, T and any space X, there is a one-to-one correspondence between the hypergesture spaces $T@(S@X^{\rightarrow})$ and $S@(T@X^{\rightarrow})$.*

3

Gestures And Content

Summary. This chapter introduces the topic of musical gestures' potential of creating contents. Gestures don't automatically wear meaning, they are not signs, such as words. Contents can however occur by gestural structures. Gestures are rather creators of contents.

– Σ –

3.1 A First Gesture for Music

Our first content producing gesture is shown in Figure 3.1.

Fig. 3.1: The simple gesture with a single arrow connects two notes, generating a melodic content.

This gesture connects two notes at different onset times and thereby generates a *melodic content*: these notes are the start and end of a melodic movement, that did not exist automatically before. In musical terms, a melodic line would mean that the two notes are played in a connected manner, legato.

G. Mazzola et al., *Basic Gestures for Music Theory, Performance, and Pedagogy*, Computational Music Science, https://doi.org/10.1007/978-3-032-10681-0_3

3.2 A Hypergesture in Basic Counterpoint

In this section we discuss a first example of the content of a hypergesture in music. We have a look at a very elementary situation in Johann Joseph Fux's famous counterpoint theory. See Figure 3.2 for the following discussion.

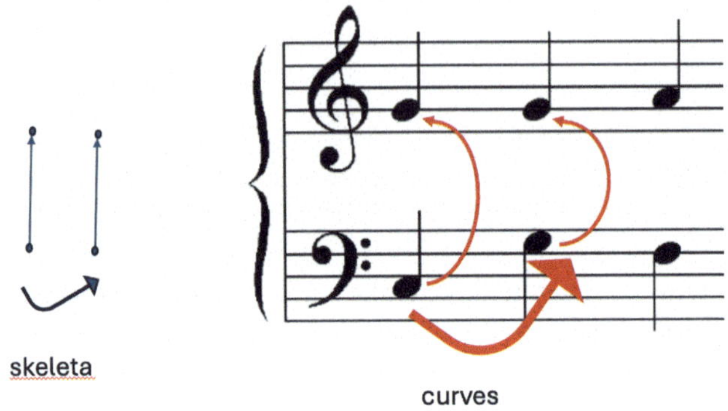

Fig. 3.2: We show a contrapuntal composition: Two voices, the lower Cantus Firmus, and the higher Discantus, put against each other in consonant intervals.

We see two interval gestures, their curves connecting the lower voice of Cantus Firmus to the higher voice of Discantus. The first interval gesture to the left is connected to the second interval gesture to the right by a hypergesture. The figure shows the skeleta of the two interval gestures (each a single arrow) as well as the skeleton of the connecting hypergesture (also a single arrow). This contrapuntal hypergesture creates the content of the Fux theory of two-voice counterpoint.

Part II

Teaching Music Theory

4

Fundamental Realities

Summary. This chapter introduces the fundamental realities, which are the basis for the development of a theory of music.

– Σ –

4.1 The Problem

The problem of musical reality is that, when we say "playing music", we understand that this playing action generates sounds in the physical reality, but the music that is played is defined in a symbolic space that exists before our performance takes place. "Playing" is a transformation, starting in a musical space, and that space is transformed or mapped into the physical reality of activated instruments, which generate waves in the air, waves, which eventually reach our ears, when we hear music's performance.

The problem of this setup resides in the *a priori* existence of a purely musical reality, which can be played, but does not depend upon such a performative transformation. In this book, we shall not delve into this philosophical aspect, but we will assume that a musical existence beyond physical performance is given. In the part of this book dealing with performance, we shall discuss the connections between musical and physical reality, i.e., the relationships of musical concepts and objects with their physical representations. In the present chapter, we ask the reader to accept the purely musical reality, together with its detailed conceptual description, which we will unfold.

This does not mean that our procedure is not precise, it will include (however simple) mathematical tools, but we will postpone their physical interpretations to the part dealing with performance. Recall in this context that the Pythagorean music theory was also independent of physical prerequisites, the Pythagorean monochord (see Figure 4.1) was only a secondary instrument to show the transcendental principle of Pythagorean music theory: the tetractys of 10 triangularly arranged points showing the octave 2:1, fifth 3:2, and fourth 4:3, see Figure 4.2.

G. Mazzola et al., *Basic Gestures for Music Theory, Performance, and Pedagogy*, Computational Music Science, https://doi.org/10.1007/978-3-032-10681-0_4

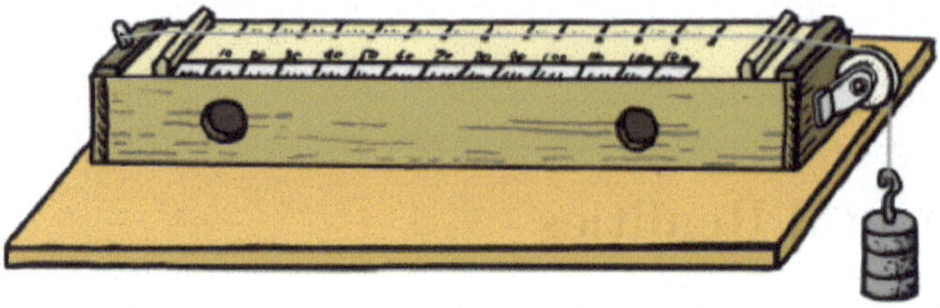

Fig. 4.1: The Pythagorean monochord.

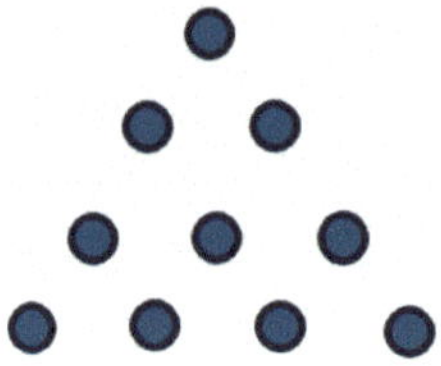

Fig. 4.2: The Pythagorean tetractys.

4.2 Creating Musical Reality

The objects of musical reality[1] are called *notes*. Music theory describes notes, or conceptual parts of notes. Here are the conceptual components of a note:

1. a note has a *pitch* p, which we describe by an integer $(\ldots -2, -1, 0, 1, 2, \ldots)$, the note's pitch can be imagined as being the key number, when the note is played on a keyboard. Following the MIDI[2] convention, on a keyboard, the middle C key is given the pitch number 60, the next higher key is 61, the next lower key is 59, etc.
2. a note also has an *onset* o, which is the concept of the time, when the note starts. The onset can be any real number, but often, the onset is described by a fraction of shape $\frac{q}{2^n}$, e.g. $\frac{1}{2}, \frac{3}{4}, \frac{12}{16}$. In Western notation these fractions represent the beats, when the notes occur.
3. a note has also a *duration* d, the time of its musical life. Like onsets, the note's duration is a real number, often of shape $\frac{q}{2^n}$, but it is always a positive number.
4. a note also shares a musical *loudness* l, which describes the loudness of the note, when played. The loudness symbols are essentially integers, but usually written in musical terms, such as $pp, p, mp, mf, f, ff, fff, \ldots$.

[1] We include several cultural contexts, but there might be music cultures, where musical reality differs from our display.

[2] MIDI = Musical Instrument Digital Interface, an industrial music standard.

5. a note also specifies the instrumental *voice* v in which it would be played. This can be described by a positive integer, 1,2,..., if we have set an enumeration of instrumental voices.
6. a note may change its pitch, and this amount is called the note's *glissando* g, it is also described by an integer, like the note's pitch.
7. lastly, a note may change its loudness, a loudness difference that is called *crescendo* c, given in similar numbers to loudness.

These seven components describe the note in its full musical reality. As already stated, we shall discuss these parameters as generators of the physical realization of a note in performance. But for the time being, we take them as basic ingredients of a standard music theory. We also write

$$n = (p(n), o(n), d(n), l(n), v(n), g(n), c(n))$$

to describe the note n by the sequence of its seven components.

It is important to understand that music theory is not always and strictly dealing with all seven note parameters. Some theoretical perspectives, such as harmony, only specify the notes' pitch, not even their loudness or duration. We must keep in mind that these perspectives are essential for understanding music theory. Music theory is the integration of all projections into partial views.

5

Elementary Concepts

Summary. This chapter introduces the elementary concepts of music theory.

– Σ –

5.1 Pitch

Fig. 5.1: A piano keyboard.

As introduced in Section 4.2, the first and theoretically most important determinant of a note n is its pitch $p(n)$, an integer. Although this is a symbolic entity, it is helpful to recognize its standard interpretation as a symbol for a potentially sounding entity. We shall make this interpretation precise in Chapter 9, but for the time being we propose an intuitive understanding. This is the interpretation of a pitch as being a number that represents, how "high" a note will sound. For example, on a piano, the note's pitch tells us where on the

G. Mazzola et al., *Basic Gestures for Music Theory, Performance, and Pedagogy*,
Computational Music Science, https://doi.org/10.1007/978-3-032-10681-0_5

keyboard to hit a key. The integer of a pitch is the number of the key, which we are hitting. As already said, the middle C key is given pitch 60. Going up on the keyboard, we get pitches 61, 62, 63, etc. Going down, we get pitch 59, 58, 57, etc., see Figure 5.1.

We say that pitch $p+1$ (resp. $p-1$) is a *semitone* higher (resp. lower) than p. The set of pitches is constructed by semitone ascending or descending steps. The set of note pitches is given an additional structure: the *octave*. Two pitches p_1, p_2 are said to be *an octave apart from each other*, if $p_2 = p_1 + 12$ or $p_1 = p_2 + 12$. Two pitches that differ by a number of octaves are said to be in the same *pitch class*. On the piano, you can hear very similar sounds from pitches in the same pitch class. The following theorem is easy to prove, but it is crucial for music theory.

Theorem 2. *If three pitches p, q, r are such that p, q and q, r are in the same pitch class, then p, r are in the same pitch class. If p, q are in the same pitch class, so is q, p. And, evidently, p is in the same pitch class as p. There are 12 pitch classes, represented by pitches* $0, 1, 2, 3, 4, 5, 6, 7, 8, 9, 10, 11$.

The last statement is proved by adding or subtracting multiples of 12 to generate pitches in the basic set $0, 1, 2, 3, 4, 5, 6, 7, 8, 9, 10, 11$. The other statements are obvious.

The set $\mathbb{Z}_{12}$ of pitch classes can be visualized as the set of hours on a circular clock. See Figure 5.2.

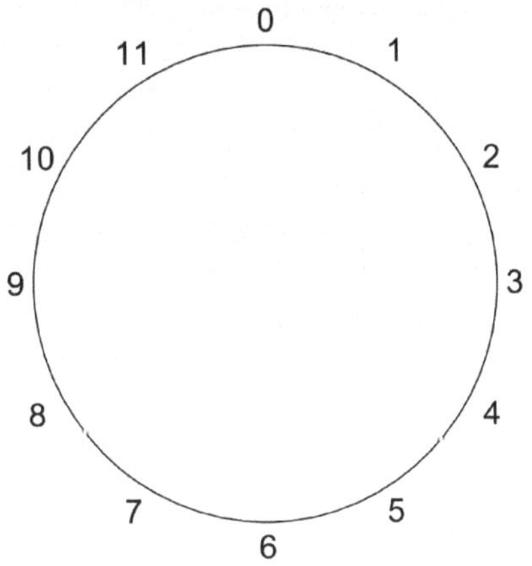

Fig. 5.2: The clock representation of pitch classes.

5.2 Pitch and Pitch Class Transformations

Pitch and pitch classes can be transformed according to a simple, but important set of transformations. If t, v are two pitches, one defines the transformation $T^t v.$ on pitches x by $T^t v.x = t + vx$, the multiplication of x by v, followed by the transposition by t. For example, for $T^8 7.$ and $x = 23$, we get $T^8 7.23 = 8+7.23 = 8+161 = 169$. If $T^t v., T^s w.$ are two transformations, their composition is again of the same structure, i.e., $T^t v \circ T^s w = T^{t+vs} vw..$ A transformation $T^t 1.$ is called a *transposition*. If two notes are in a semitone distance from each other, multiplication by 7 generates two notes which are a fifth, i.e., 7 semitones apart from each other.

Such transformations can also be defined on pitch classes as follows: One may define sum and product of pitch classes c, d by taking the sum and product of pitches that define these classes and then taking the pitch classes of these operations. For example, if c, d are representing pitches p_c, p_d we have $c + d$ the pitch class of $p_c + p_d$, and $c.d$ the pitch class of $p_c.p_d$. For example, if c, d are represented by $5, 7$, the product is the pitch class of $5.7 = 35$, which is the class of 11. This has four remarkable special results: for pitch classes $p = 5, p = 7, p = 1, p = 11$, their square is pitch class 1, since $5.5 = 7.7 = 1.1 = 11.11 = 1$. Other pitch classes behave differently, e.g., $3.4 = 12 = 0$. On the set of pitch classes, when represented by points on the circle, as in Figure 5.2, a transposition by t is given by a rotation on the circle by t pitch class units.

5.3 Chords and Scales

The prominent objects of harmony are *chords*, which are defined as collections of pitch classes with 3 or more elements. Collections of two pitch classes are called *intervals*. Prominent intervals are transpositions of *fifth intervals* $5th = \{0, 7\}$, *fourth* intervals $4th = \{0, 5\}$ *minor second* intervals $min2 = \{0, 1\}$, and *major second* intervals $maj2 = \{0, 2\}$.

If ch is a chord, the set of all pitches that project into ch is called the chord's *scale*. Chords, whose scale is interesting are also called *scale-chords*. Chords with three elements are called *triads*. There are these types of important triads:

1. *major* triads, they are all transpositions of the triad $Major = \{0, 4, 7\}$;
2. *minor* triads, they are all transpositions of the triad $Minor = \{0, 3, 7\}$;
3. *dimished* triads, they are all transpositions of the triad $Dim = \{0, 3, 6\}$;
4. *augemented* triads, they are all transpositions of the triad $Aug = \{0, 4, 8\}$.

And here are some important scale chords:

1. *major scale* chord $Maj = \{0, 2, 4, 5, 7, 9, 11\}$;
2. *melodic minor scale* chord $Mel = \{0, 2, 3, 5, 7, 9, 11\}$;
3. *harmonic minor scale* chord $Harm = \{0, 2, 3, 5, 7, 8, 11\}$;

4. *whole tone scale* chord $Whole = \{0, 2, 4, 6, 8, 10\}$;
5. *tetrade complement scale* chord $Tetra = \{0, 1, 3, 4, 6, 7, 9, 10\}$, complement of tetra chord $\{2, 5, 8, 11\}$;

Music theory often considers a combination of different chords as its relevant objects. In the following section we discuss some prominent combinations.

5.3.1 Tonalities

On the most elementary conceptual level, a *tonality* is a pair (S, t) of a scale chord S and an element $t \in S$ [1], which is called the *tonic* of S. The most famous example is the tonality of C Major, which is defined by the pair $(Maj, 0)$. One may apply a transposition T^t to C Major, getting the tonaliy $(T^t.Maj, t)$, also called *t Major.* If we vary the tonic in the scale chord Maj, we get the so-called modes, here are their names (the scale Maj being fixed):

1. tonic = 0 defines C Major or *ionian* mode;
2. tonic = 2 defines *dorian* mode;
3. tonic = 4 defines *phrygian* mode;
4. tonic = 5 defines *lydian* mode;
5. tonic = 7 defines *mixlydian* mode;
6. tonic = 9 defines *aeolian* mode;
7. tonic = 11 defines *locrian* mode.

Another contribution to the concept of a tonality is the set of the above defined triads $tr \subset S$ [2]. For the major tonality, C Major, say, we exhibit seven triads as follows, and they are called *degrees*:

1. degree $I = \{0, 4, 7\}$, a major triad;
2. degree $II = \{2, 5, 9\}$, a minor triad;
3. degree $III = \{4, 7, 11\}$, a minor triad;
4. degree $IV = \{5, 9, 0\}$, a major triad;
5. degree $V = \{7, 11, 2\}$, a major triad;
6. degree $VI = \{9, 0, 4\}$, a minor triad;
7. degree $VII = \{11, 2, 5\}$, a diminished triad.

For transposed major scales, the degrees are defined by the evident transposition of the C Major degrees.

The configuration of the seven triadic degrees has a wonderful geometric description, which we want to construct now. The geometry starts with drawing all seven degrees as points $I, II, III, IV, V, VI, VII$ in the normal three-dimensional space, see Figure 5.3.

Next, we connect any two degree points with a line if they have pitch classes in common, see Figure 5.4

---

[1] $\in$ = element

[2] $\subset$ = subset

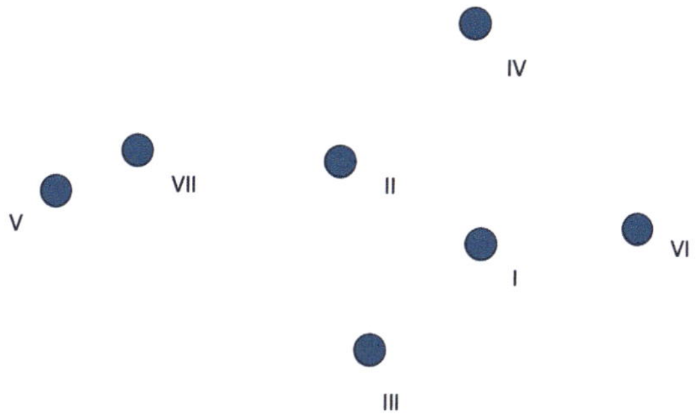

Fig. 5.3: Drawing the seven triadic degrees as points in space.

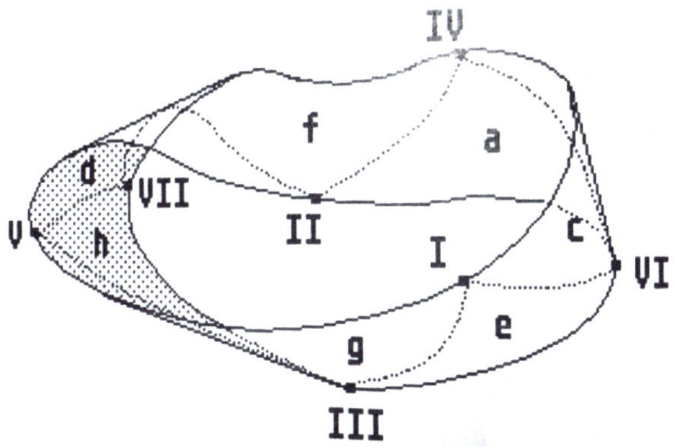

Fig. 5.4: Drawing the lines connecting any two (different) degree points that have pitch classes in common

This line diagram was conceived by composer Arnold Schoenberg in his harmony book [93], he called the lines "harmonic bands", but he did not draw this configuration in a space, like we do.

The next step is a dramatic one, and Schoenberg was not aware of its possibility. In this step, we look for any three degrees, which have a pitch class in common. For example, degrees I, III, V have the pitch class 7 in common. For any three degrees that have common pitchclasses, we draw a triangular surface spanned by these three degrees. What we get is shown in Figure 5.5.

The result is remarkable: the generated surface is a Moebius strip, like a belt that has been connected in a wrong way. This geometric structure was unknown until this author, Guerino Mazzola, discovered it in 1985 [58]. We shall give a music-theoretical interpretation soon. Essentially, we shall learn that the Moebius geometry is responsible for major problems of traditional harmony that was construed by Hugo Riemann.

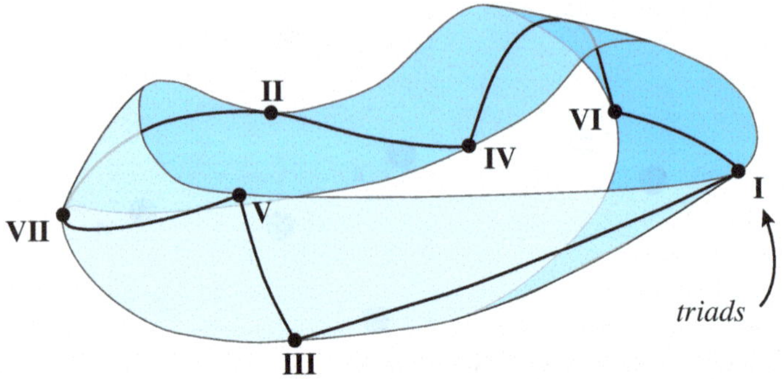

Fig. 5.5: Drawing the triangular surfaces connecting any three (different) degree points that have pitch classes in common. This generates a Moebius strip, a belt that is connected in "wrong" way.

It is remarkable that this construction that connects degrees can be interpreted as a gesture, whose skeleton is the system of 7 points and their connecting arrows. This gestural interpretation of the Moebius strip gives the construction a new perspective, which was not present in the original approach.

5.3.2 Riemann's Failure

The Moebius strip has dramatic consequences for Hugo Riemann's approach to harmony. For Riemann, a specific tonality meant to define a function $R : CHORDS \to \{T, D, S\}$ on the set $CHORDS$ of all chords with three possible values, T for Tonic, D for Dominant, S for Subdominant. This program failed for specific reasons relating to such a function on the standard triads on a major scale. We will explain this now.

Riemann asked that his function R on the border of the Moebius surface must be defined such that R's values would be different for successive triads. For example, $R(V)$ should differ from $R(I)$. Second, standing on the strip's surface, taking on a triad to the left of the surface, e.g., on V, and looking in direction the border, e.g., to I from the V, there is a specific triad to the right in front of this position, e.g. looking to III from V. Riemann asks for the R function to be the same for III and for V, III is called the *parallel* triad to V. The problem is that because of the Moebius geometry, standing on III on the back side of the strip, with the surface to your right, the parallel to III is I. Therefore $R(V) = R(III)$ and $R(III) = R(I)$, whence $R(V) = R(I)$, which contradicts Riemann's condition that $R(V)$ and $R(I)$ should be different.

This means that *the Moebius gestural geometry of triads is responsible for Riemann's failure in harmony.* This is a new perspective on harmonic logic.

5.3.3 Chord Symmetries

In this section, we want to focus on transformations that act upon chords, in particular: transformations $s = T^t v$ (on an argument x, this yields $T^t v.x = t + v.x$) on the pitch class set $\mathbb{Z}_{12}$, which leave invariant a determined chord ch, i.e., $s.ch = ch$. This topic will be important to model the transition from one tonality to another one, a procedure that is known as *tonal modulation*, and which we shall discuss in Chapter 6.

The most prominent symmetries are: transposition T^t by pitch class t, e.g. $T^t.7.1 = t + 7$, inversion $T^t 11$, e.g. $T^t 11.5 = t + 11.5 = t - 5$, or $T^{2t} 11.t = t$. The inversion has this fixpoint: $T^t 11.s = s$, if $t = 2s$. It has no fixpoint if t is not even. Transpositions have $T^t.s = s$ iff $t + s = s$, $t = 0$ only. Figure 5.6 shows some such symmetries.

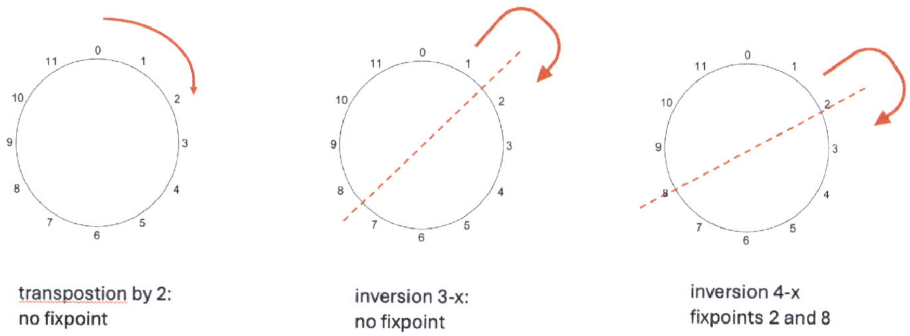

Fig. 5.6: Some transpositions and inversions, as shown on the $\mathbb{Z}_{12}$ clock.

Here are some symmetries s for important chords ch, i.e., symmetries s that leave chords invariant: $s.ch = ch$.

1. major and minor triads have no symmetries(, except identity);
2. diminished and augmented triads have a single inversion through their middle as symmetries, e.g., $ch = \{0, 3, 6\}$ has the inversion symmetry $s = T^6.11$ around 3.
3. the major scale C Major has the inversion $T^4 11$ around its pitch class 2 as symmetry; this can be visualized on a piano keyboard, which has a symmetry around the key d.
4. the melodic minor scale has inversion $T^2 11$ around 1 as unique symmetry;
5. the harmonic minor has no symmetry;
6. the whole tone scale has inversions around 0 and 1 plus transpositions by 2 as its basic symmetries;
7. the tetrade complement scale has inversions around 1/2 and around 2 plus transposition by 3 as its basic symmetries.

5.3.3.1 Symmetries are Gestures

Symmetries are conceptually different from gestures. But in our musictheoretical context, we might reinterpret symmetries on $\mathbb{Z}_{12}$ as gestures by the following interpretation, see Figure 5.7.

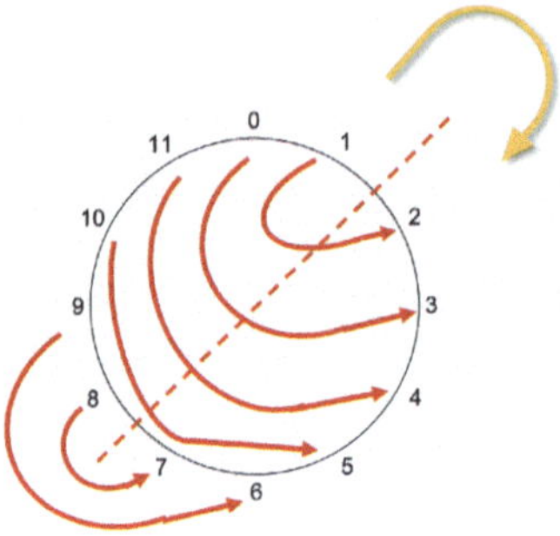

Fig. 5.7: A symmetry on $\mathbb{Z}_{12}$ reinterpreted as a gesture.

The gestural interpretation is enabled, when we extend the clock $\mathbb{Z}_{12}$ as being a subset of the plane, where we draw it. In this view, points of the clock become points in 2-space, the plane, where it is drawn. In this perspective, a symmetry s connects two points, the argument x on the clock, to its symmetry image $s.x$. This connection can be visualized by a gestural curve from x to $s.x$, as shown in Figure 5.7. This creates a gesture $G(s)$ for symmetry s, where its curves move out of the clock, when connecting its points. This *embedding of the clock in the plane* will be used when constructing models of tonal modulation (change of tonality).

5.3.3.2 Cadences

To identify tonalities, one looks for collections of triadic degrees, which uniquely determine the tonalities.

Definition 1 *For a given major tonality t, a set c of triads in t is called a cadence of t, if no other tonality within a given collection of tonalities shares that set c.*

Among all cadences for t, there are minimal such sets. Here are all minimal cadences:

1. $c = \{II, III\}$
2. $c = \{II, V\}$
3. $c = \{III, IV\}$

4. $c = \{IV, V\}$
5. $c = \{VII\}$

To understand these five options, we look at a particular transformation of the major scale $t = \{0, 2, 4, 5, 7, 9, 11\}$, namely $t' = T^1 7.t = \{0, 1, 2, 3, 4, 5, 6\}$, which is a full sequence of points on the clock between 0 and 6. This is a consequence of the fact that a major scale is a sequence of fifth (7) intervals: f-c-g-d-a-e-b. The cadences are constructed by the condition that in the t' presentation, we have to show the border points $0, 6$ in the cadential triads. To include 0, we need triads II, IV, VII, for 6, we need VII, V, III. For VII we don't need another triad. Else, we need to pair II, IV with V, III, which yields the 2 x 2 remainder cadences.

6

Melodies, Harmonies, Modulation, Counterpoint

Summary. In this chapter we discuss important parts of music theory: melodies, harmonies, modulation, and counterpoint

– Σ –

6.1 Melodies and Harmonies

In music theory, they use two special sets of pitch classes: melodies and chords. We have introduced chords in the previous chapter. We shall learn how chords are used in the theory of transitions from one tonality to another in Section 6.2. We shall discuss a theory of melodies, Johann Joseph Fux's counterpoint in Section 6.3.

6.2 Modulation

In this section we want to describe a harmonic procedure as initiated by Arnold Schönberg in his harmony book [93]. The idea is that we first play in a major scale t and then play a transitory phase which leads us to a second major scale s. In Schönberg's logic there are three phases: phase 1 plays triads of t which neutralize t, i.e., we play triads that show non-cadential triads of t. Phase 2 introduces so-called *pivotal* degrees of the addressed scale s, degrees which represent the movement to the new tonality s. Phase 3 introduces a cadence c of s to identify s. For example, going from t to $s = T^5 1.t$, the fourth transposition of t, Schönberg's logic requires the pivotal degrees II, IV, VII of s.

We will now describe a system of transitions, which coincides with Schönberg's rules for any two major tonalities t, s.

Given the pair t, s, $s = q.t$, q *the modulator symmetry*, and a minimal cadence $m \subset s$, a *modulation quantum* M from t to s is a subset $M \subset \mathbb{Z}_{12}$ with the following properties

G. Mazzola et al., *Basic Gestures for Music Theory, Performance, and Pedagogy*, Computational Music Science, https://doi.org/10.1007/978-3-032-10681-0_6

- the modulator q is a symmetry of the M;
- the triads of m are subsets of M;
- the symmetries $Sym(M)$ of M have a trivial intersection with $T^{\mathbb{Z}_{12}} \pm 1$;
- $s \cap M^1$ is covered by triads of s;
- M is minimal with the first two above properties.

The following is the theorem of quantized modulations:

Theorem 1 *For any combination t, s, with $s = q.t$, q the modulator, there is a modulation quantum M with a minimal cadence m. The triads of s in M coincide with Schönberg's pivotal degrees wherever he had defined such a structure.*

We exemplify this conceptualization by a modulation C Major to F Major, as shown in Figure 6.1 with a score representation.

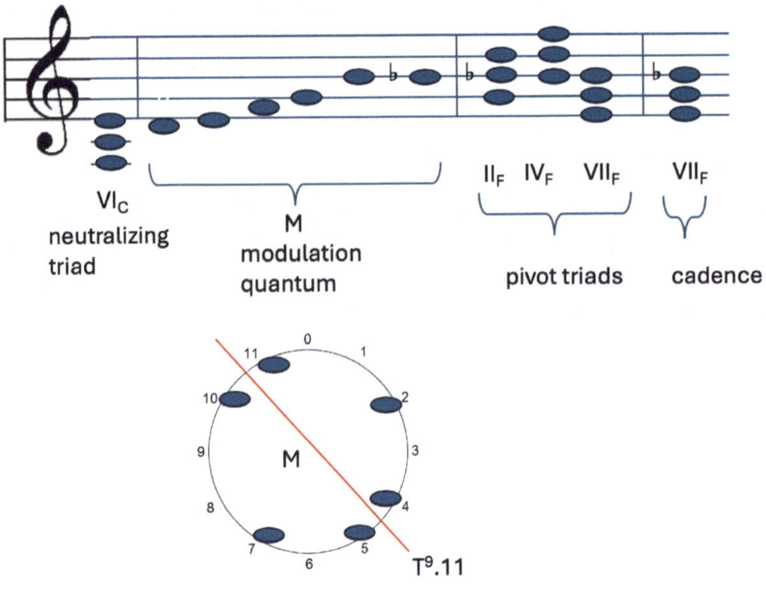

Fig. 6.1: The modulation components from C Major to F Major. To the left a neutralizing triad VI_C, in the middle the modulation quantum M together with the symmetry $T^9 11$ mapping C Major to F Major. To the right the pivotal F-triads II, IV, VII and the cadence VII in F Major.

Figure 6.2 shows all possible cases, together with the geometric degree intersections. E.g., for one fourth transition (e.g. C Major $\rightarrow$ F Major) , we see the three pivotal degrees II', IV', VII' in the Figure's left middle graphic $1/5$.

[1] $\cap$ = intersection

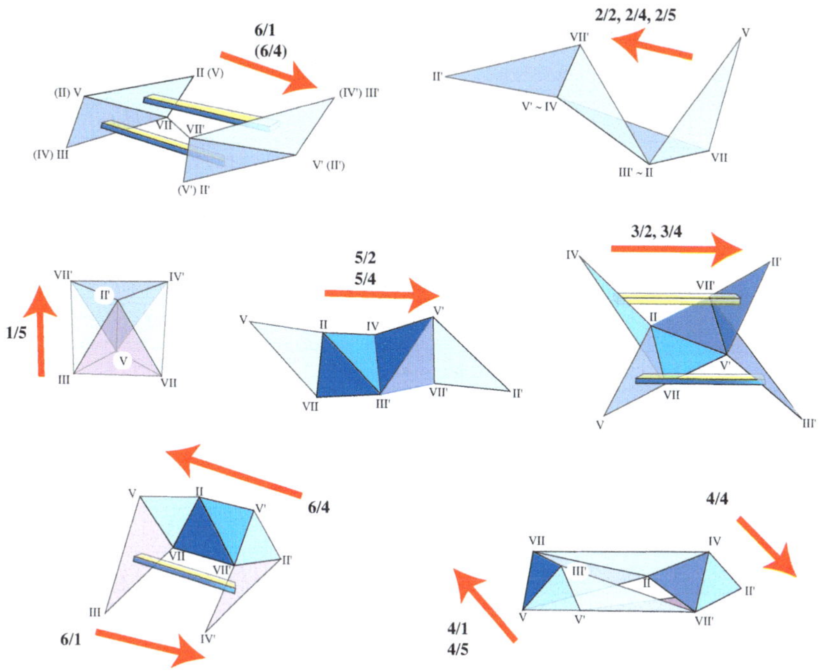

Fig. 6.2: All modulation quanta of the major scale.

6.2.1 Quanta and Pivots for the Modulations Between Diatonic Major Scales

Quanta and Pivots for the Modulations Between Diatonic Major Scales				
Transl. p	*Cadence*	*Quantum*	*Modulator*	*Pivots*
1	$\{II, V\}$	● ○ ● ● ○ ● ● ● ● ● ● ●	$T^5 11$	$\{II, III, V, VII\}$
1	$\{II, III\}$	● ○ ● ● ○ ● ● ● ● ● ● ●	$T^5 11$	$\{II, III, V, VII\}$
2	$\{VII\}$	○ ● ● ○ ● ● ○ ● ○ ○ ○ ●	$T^6 11$	$\{II, IV, VII\}$
2	$\{II, V\}$	○ ● ● ○ ● ● ○ ● ○ ● ○ ●	$T^6 11$	$\{II, IV, V, VII\}$
2	$\{IV, V\}$	○ ● ● ○ ● ● ○ ● ○ ● ○ ●	$T^6 11$	$\{II, IV, V, VII\}$
3	$\{II, V\}$	● ○ ● ○ ○ ● ○ ● ● ● ● ●	$T^7 11$	$\{II, III, V, VII\}$
3	$\{II, III\}$	● ○ ● ○ ○ ● ○ ● ● ● ● ●	$T^7 11$	$\{II, III, V, VII\}$
4	$\{VII\}$	○ ○ ● ● ○ ● ● ○ ○ ● ○ ●	$T^8 11$	$\{II, IV, V, VII\}$
4	$\{IV, V\}$	○ ● ● ● ● ● ● ● ○ ● ○ ●	$T^8 11$	$\{II, III, V, VII\}$
4	$\{II, III\}$	● ● ● ● ○ ● ● ● ● ● ○ ●	$T^8 11$	$\{V, VII\}$

Quanta and Pivots for Diatonic Major Scales—Continued

Class Transl. p	*Cadence*	*Quantum*	*Modulator*	*Pivots*
5	$\{VII\}$	○○●○●●○●○○●●	$T^9 11$	$\{II, IV, VII\}$
6	$\{II, III\}$	○●●●●●○●●●●●	T^6	$\{II, III, V, VII\}$
6	$\{IV, V\}$	○●●●●●●●●●○●	$T^{10} 11$	$\{II, IV, V, VII\}$
6	$\{IV, V\}$	●●●●○●●●●●○●	T^6	$\{II, IV, V, VII\}$
6	$\{II, III\}$	●●●●○●○●●●●●	$T^{10} 11$	$\{II, III, V, VII\}$
7	$\{VII\}$	●○●○○●●○○●○●	$T^{11} 11$	$\{III, V, VII\}$
8	$\{VII\}$	○●●○○●○●○○●●	$T^0 11$	$\{II, VII\}$
8	$\{IV, V\}$	○●●●●●○●●●●●	$T^0 11$	$\{II, IV, V, VII\}$
8	$\{II, III\}$	●●●●○●○●○●●●	$T^0 11$	$\{II, III, V, VII\}$
9	$\{II, V\}$	○○●○●●●●●●○●	$T^1 11$	$\{II, IV, V, VII\}$
9	$\{IV, V\}$	○○●○●●●●●●○●	$T^1 11$	$\{II, IV, V, VII\}$
10	$\{VII\}$	●○●●○●○○○●○●	$T^2 11$	$\{III, V, VII\}$
10	$\{II, V\}$	●○●●○●○●○●○●	$T^2 11$	$\{II, III, V, VII\}$
10	$\{II, III\}$	●○●●○●○●○●○●	$T^2 11$	$\{II, III, V, VII\}$
11	$\{II, V\}$	○●●○●●●●●●●●	$T^3 11$	$\{II, IV, V, VII\}$
11	$\{IV, V\}$	○●●○●●●●●●●●	$T^3 11$	$\{II, IV, V, VII\}$

This quantitative and qualitative display of modulations among major scales also plays out for all modes that are subsisting with the defining major scales. It is the most precise and comprehensive modulation theory, and beyond confirming Schönberg's approach, it realizes a mathematically explicit model. This model also comprises other scales, see [62, Chapter 27].

Figure 6.3 shows an example of all modulations.

6.2.2 A Second Modulation Model, which is Based on Gestures

The previous modulation model uses symmetries between modulating tonalities. Although such symmetries have been recognized as being gestures, it is also possible to model tonal modulation starting directly from gestures. We want to sketch this second model now, and we can state that the result is the same, i.e., the pivot degrees coincide with those obtained in the symmetry-oriented model.

We shall describe the modulation from C Major to F Major. We represent the C major scale starting with d on top, going down to the right to $d\#$, e, f, $f\#$, g, $g\#$, a, $a\#$, b, c, $c\#$, returning to the initial d. The symmetry of the

Fig. 6.3: This score shows a sequence of modulations into all major tonalites. The tonalities are in German notation, where B stands for B flat, and H stands for B.

C scale is now represented by a vector field $X(x, y) = (cos(y)cos(x), 0)$ on the plane that comprises the clock, see left part of Figure 6.4.

The corresponding symmetry of F Major is shown to the right in Figure 6.4. Using a famous operator in mathematical physics, the *Lie bracket*, we construct a new vector field $[X, Y]$, Lie bracket of X and Y. Figure 6.5 shows this Lie bracket to the left, also showing the clock circle that is embedded in the Lie bracket plane. To the right of this figure, one shows integral curves with their intersection with the clock. We recognize that three integral curves intersect with all 12 clock points, which is a miraculous fact.

The three Lie bracket integral curves enable hypergestural deformations of triadic gestures, as shown in Figure 6.6

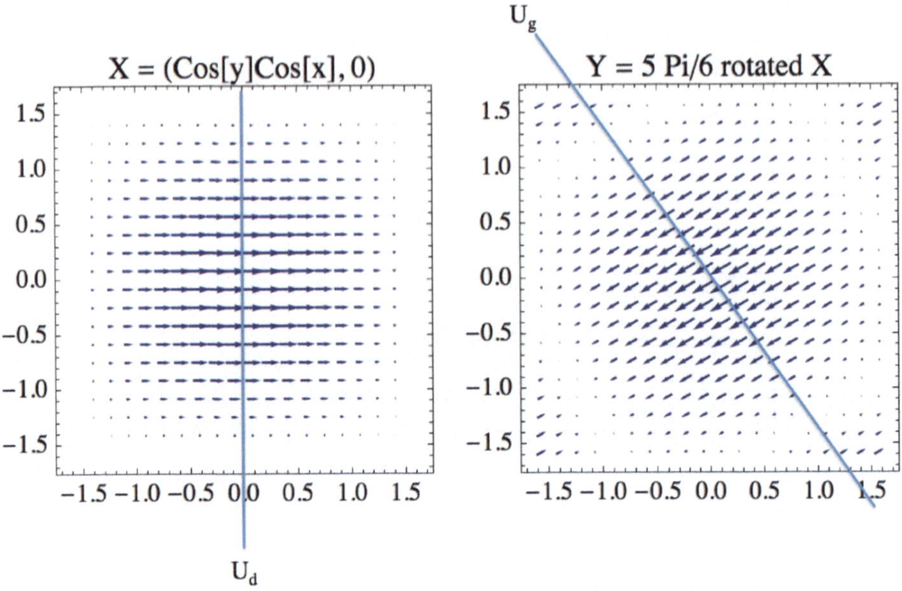

Fig. 6.4: Left:The vector field X for C Major, right: the corresponding vector field Y for F Major. These vector fields show arrows on the 2D plane, which are defined by the formuas above the graphic representations.

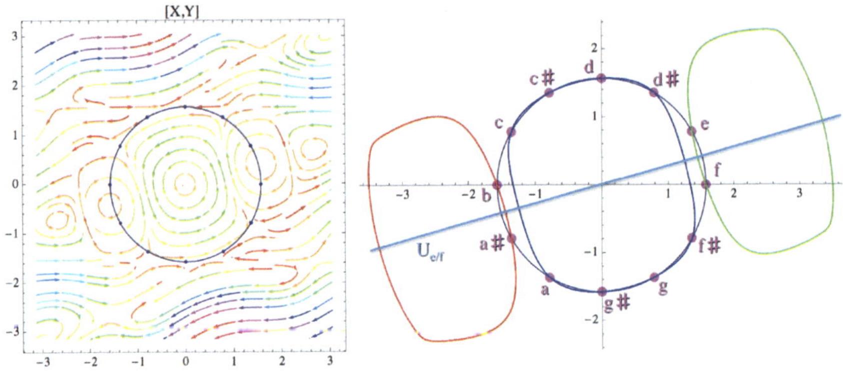

Fig. 6.5: Left:The Lie bracket vector field, right: the three integral curves, including all 12 points of the clock.

It can be shown (see [64, Chapter 80]) that with this hypergestural deformation technique, one may produce the classical Schönberg pivotal degrees, and this again for all pairs of major tonalities.

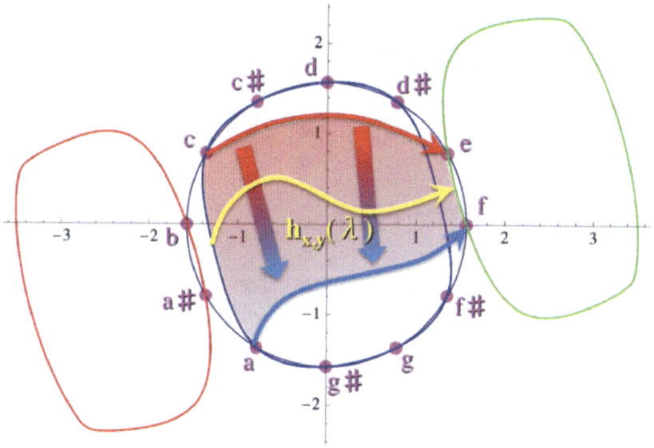

Fig. 6.6: A gesture from c to e, in fact a part of degree I in C Major, is deformed to gesture from a to f, part of the triadic degree I in F major.

6.3 Counterpoint

Counterpoint is a theory, having been developed in the Middle Ages and perfectioned by Johann Joseph Fux in his book *Gradus ad Parnassum* [20, 1725], which is still taught in our days! Counterpoint describes the succession of musical intervals in time. So, the objects of counterpoint are intervals, i.e., by definition pairs (CF, D) (CF = cantus firmus, D = discantus) of pitches of such intervals, together with sequences $(CF_1, D_1), (CF_2, D_2), \ldots, (CF_n, D_n)$ of such intervals with increasing onset time. Figure 6.7 shows such a sequence. The time interval between successive intervals is usually such that the difference of onsets is the duration of the first interval pitches. Counterpoint theory describes and calculates the allowed intervals that follow a given interval. In this section, we shall display a theory for such allowed interval sequences.

The basic fact for counterpoint is the existence of a unique *autocomplementary* transformation $p : Consonances \to Dissonances : x \to p(x) = 5x+2$, which connects the consonance set $Consonances = \{0, 3, 4, 7, 8, 9\}$ to the set $Dissonances = \{1, 2, 5, 6, 10, 11\}$.

This transformation, called *strong*, is applied to the following space of intervals: It is the space $\mathbb{Z}_{12} + \epsilon.\mathbb{Z}_{12}$. Its elements $x + \epsilon.y$ (cantus firmus x and interval y) can be added and multiplied as follows: $x + \epsilon.y + u + \epsilon.w = (x+u)+\epsilon(y+w)$, $(x+\epsilon.y).(u+\epsilon.w) = (x.u)+\epsilon(x.w+y.u)$. An element $x+\epsilon.y$ represents the interval from cantus firmus x and interval quantity y, yielding the discantus pitch $x + y$.

The symmetry on $\mathbb{Z}_{12} + \epsilon.\mathbb{Z}_{12}$ that corresponds to p is $T^{(x+\epsilon.2)}5$, it yields $T^{x+2\epsilon}5(u+\epsilon.v) = x+5u+2\epsilon+5.v.\epsilon$. If v is a consonance, we get $x+5u+(2+5.v)\epsilon$, a dissonance at cantus firmus $x + 5u$. In other words, p transforms consonant intervals $u + \epsilon.v$ to dissonant intervals $x + 5u + \epsilon.(2 + 5.v)$ at $x + 5u$.

Applying this symmetry to $x + 5u + \epsilon.(2 + 5.v)$ yields $6x + u + \epsilon.v$, the original consonance at cantus firmus $6x + u$. This means that p induces a one-

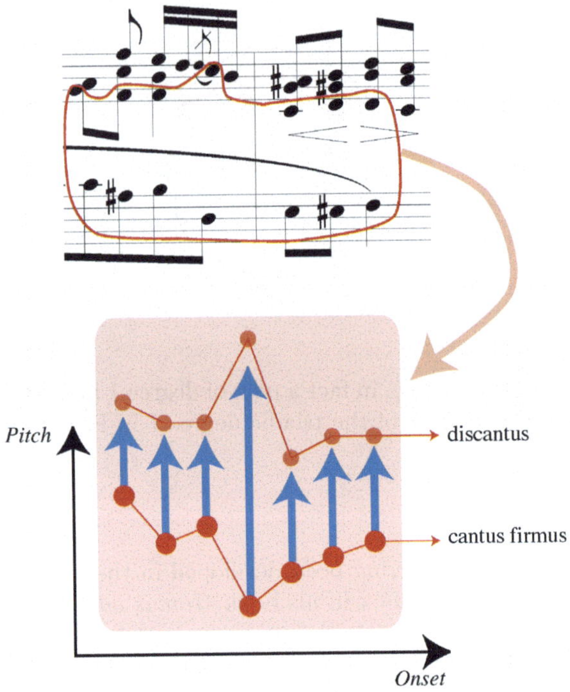

Fig. 6.7: A sequence of intervals for counterpoint theory.

to-one transformation of consonant intervals to dissonant intervals of a different cantus firmus.

On the consonance-dissonance splitting Δ on $\mathbb{Z}_{12} + \epsilon.\mathbb{Z}_{12}$, there is a unique *autocomplementary* transformation p^x_Δ (exchanging consonances and dissonances) such that the "tangent space" I_x of intervals with cantus firmus x remains fixed. On I_x, p^x_Δ acts in the same way as the original p.

Next, we introduce contrapuntal symmetries:

Definition 2 *If Δ is the dichotomy of consonances and dissonances, and if $\xi = x + \epsilon.i \in Consonances$, a symmetry g is called* contrapuntal *for ξ if $\xi \notin g(Consonances[\epsilon])$; p^x_Δ is a polarity (exchanges $g.Consonances$ and $g.Dissonances$) of $g.\Delta$; and the cardinality of $g(Consonances[\epsilon]) \cap Consonances[\epsilon]$ is maximal with the first two properties.*

The relationship between an interval and a possible successor is described in following proposition:

Theorem 2 *If we are given the strong dichotomy*

$$\Delta = Consonances[\epsilon]/Dissonances[\epsilon],$$

then, for a given interval ξ, *an interval* η *is a* successor *of* ξ *if it is contained in the intersection* $g(Consonances[\epsilon]) \cap Consonances[\epsilon]$ *for a contrapuntal symmetry* g.

The following table describes all successors in the major scale.

1. Oblique Motion in Cantus Firmus						
$CF: 0 \mapsto 0$	$CF: 2 \mapsto 2$	$CF: 4 \mapsto 4$	$CF: 5 \mapsto 5$	$CF: 7 \mapsto 7$	$CF: 9 \mapsto 9$	$CF: 11 \mapsto 11$

0 4 7 9
    * * 0
*   * * 4
* *   * 7
* * *   9
	0 3 7 9
	  * * * 0
	* * 3
	* * * 7
	* * 9
		0 3 7 8
		  * * * 0
		* * * 3
		* * * 7
		* * * 8
			0 4 7 9
			    * * 0
			* * * 4
			* * * 7
			* * * 9
				0 4 7 9
				    * * 0
				* * * 4
				* * * 7
				* * * 9
					0 3 7 8
					  * * * 0
					* * * 3
					* * * 7
					* * * 8
						0 3 8
						  * * 0
						* * 3
						* * 8

2. Minor Ascending Second in Cantus Firmus						
$CF: 4 \mapsto 5$	$CF: 11 \mapsto 0$					
0 3 7 8 * * * * 0 * * * * 4 * * * 7 * * * * 9	0 3 8 * * * 0 * * * 4 * * * 7 * * * 9					
3. Minor Descending Second in Cantus Firmus						
$CF: 5 \mapsto 4$	$CF: 0 \mapsto 11$					
0 4 7 9 * * * * 0 * * * * 3 * * * 7 * * * * 8	0 4 7 9 * * * * 0 * * * * 3 * * * * 8					
4. Major Ascending Second in Cantus Firmus						
$CF: 0 \mapsto 2$	$CF: 2 \mapsto 4$	$CF: 5 \mapsto 7$	$CF: 7 \mapsto 9$	$CF: 9 \mapsto 11$		
0 4 7 9 * * * 0 * * * * 3 * * * 7 * * * * 9	0 3 7 9 * * * * 0 * * * * 3 * * * 7 * * * * 8	0 4 7 8 * * * 0 * * * 4 * * * 7 * * * * 9	0 4 7 9 * * * 0 * * * * 3 * * * 7 * * * 8	0 3 7 8 * * * * 0 * * * * 3 * * * * 8		
5. Major Descending Second in Cantus Firmus						
$CF: 2 \mapsto 0$	$CF: 4 \mapsto 2$	$CF: 7 \mapsto 5$	$CF: 9 \mapsto 7$	$CF: 11 \mapsto 9$		
0 3 7 9 * * * 0 * * * * 4 * * * 7 * * * * 9	0 3 7 8 * * * * 0 * * * * 3 * * * 7 * * * * 9	0 4 7 9 * * * 0 * * * 4 * * * 7 * * * * 9	0 3 7 8 * * * 0 * * * * 4 * * * 7 * * * 9	0 3 8 * * * 0 * * * 3 * * * 7 * * * 8		

Observe that a fifth interval $x + \epsilon.7$ can never be followed directly by another fifth interval. This is the rule of forbidden parallels of fifths.

It is remarkable that our strong dichotomy of consonances vs. dissonances is one of a total of six strong dichotomies, and one may develop a counterpoint for each of the six dichotomies, see [7]. Here is the complete list:

1. $(\{1, 2, 3, 4, 5, 11\} | \{0, 6, 7, 8, 9, 10\})$

2. $(\{0,2,7,8,9,11\}|\{1,3,4,5,6,10\})$
3. $(\{0,1,2,6,7,9\}|\{3,4,5,8,10,11\})$
4. $(\{0,2,4,7,8,11\}|\{1,3,5,6,9,10\})$
5. $(\{0,2,4,6,7,10\}|\{1,3,5,8,9,11\})$
6. the consonance-dissonance dichotomy

7

Gestures In Theory And Scores

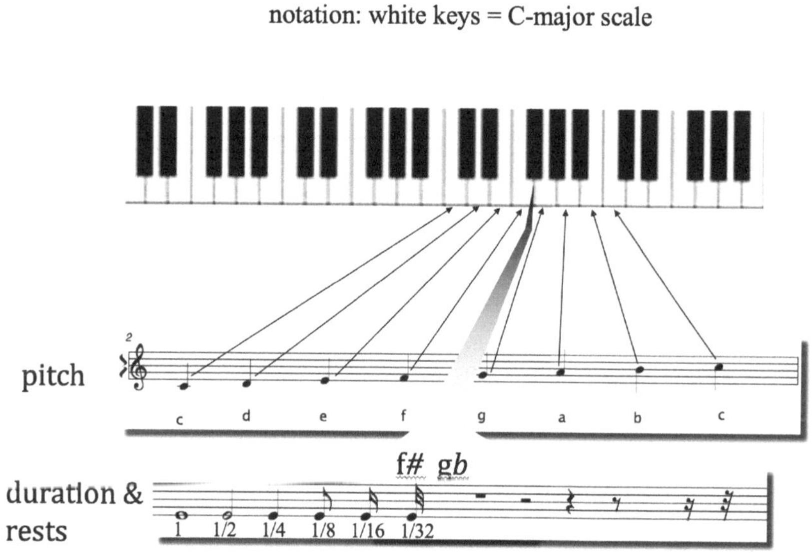

Fig. 7.1: The blank part represents the black key to the right of the f key, it is the f♯ pitch, or else (same content) the $g\flat$ pitch.

Summary. The music theory, which was presented in the previous chapter dealt with musical objects, such as pitches, pitch classes, and note abstractions focusing on pitch and onset and duration. But in music, such abstractions are always referring to compositions. In this chapter, we want to construe the standard environment, where compositions are living, and where music theory applies, called the score. Be aware that this is still a musical reality, which is de-

G. Mazzola et al., *Basic Gestures for Music Theory, Performance, and Pedagogy*,
Computational Music Science, https://doi.org/10.1007/978-3-032-10681-0_7

tached from performance in the physical reality; we shall deal with performance in the next part of this book.

– Σ –

7.0.1 Duration of Notes and Pauses

Notes' durations are symbolized by special graphical symbols, also including pauses' duration symbols, see Figure 7.1

A score (of a musical composition) is a graphical representation of notes to be played. The pitches are shown in Figure 7.1 by horizontal lines, the pitches are given by positions on or between the lines, the *g* pitch above middle *c* is encircled by the treble clef's lines. Although pitch normally is an integer, the glissando enables a smooth transition between two pitches, extending the pitch numbers to real numbers, i.e., all decimal numbers between the starting and ending pitches, as shown in Figure 7.2.

Fig. 7.2: Glissando is the smooth shift from an initial pitch to a final pitch.

More generally, the embedding of note parameters in real number domains enables gestural movements in scores, as also shown in Figure 7.3

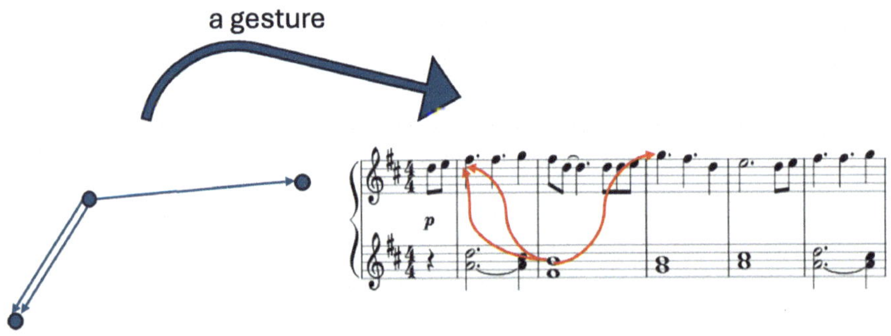

Fig. 7.3: Gestures in real number spaces for musical scores.

Part III

Performance

8

Definition of Performance

Summary. Performance theory describes the transformation process from the symbolic music representation to the physical embodiment of *played* music in a collection of sounds.

– Σ –

8.1 The Historical Unfolding of Performance Theory

Originally, performance of music in Europe was a secondary option, the Christian approach was a representation of symbolic entities aiming at God's higher reality, to be represented by choirs of singers. This approach became the topic of philosophical reflections focusing on the deeper meaning of making music.

A first look at the history of performance research seems to confirm Reinhard Kopiez's thesis [40] that there are two unrelated threads: on the one hand the empirical research as typically represented by Carl Seashore's measurements of agogics [40], and on the other the philosophical research as typically represented by Theodor Wiesengrund Adorno's writings [3]. It is true that these threads can be recognized, and we shall give their short history below, but it is erroneous to consider them as being in any contradictory position. The difference of these approaches to performance is this: The empirical research deals with quantitative aspects of performance, with the numerical specification of all the parameters that are necessary to describe the structure of a performance. The philosophical approach is concerned with the qualitative aspects of performance, with the thoughts that might be expressed through performance, i.e. the semantic content of such expressivity.

The apparent contradiction of these positions stems from the difficult and long unsettled problem of how to connect measurable quantities to thoughtful qualities in the context of a valid theory. This means that the positions are not in contradiction, but only disconnected from each other by a historical lack of links, which could generate quantitative correspondences to semantic

G. Mazzola et al., *Basic Gestures for Music Theory, Performance, and Pedagogy*,
Computational Music Science, https://doi.org/10.1007/978-3-032-10681-0_8

qualities. It is also true that empirical measurements could not be compared to precise predictions issued from theoretical models, either because the models were strictly qualitative, or because their predictive power was not precise enough. Finally, the early discussions of performance were not driven by a scientific spirit, but by a pronouncedly practical perspective driven by performance pedagogy.

In the following sections, we shall first describe this philosophical perspective. We then develop the theory of performance as a mathematically structured mapping from symbolic reality into physical reality.

8.1.1 The Philosophical Performance Theory

For a rather thorough account on the philosophical tradition, we refer to Hermann Danuser's excellent report *Musikalische Interpretation* [16], the only major omission being an adequate account on Hugo Riemann's contribution to performance theory Musikalische Dynamik und Agogik. [87].

Danuser exhibits a triple thematization of reflection on performance. Although one can never claim a disjoint splitting of a history of ideas, we can schematically depict this tripartition as follows: The historical development exhibits a first phase, roughly lasting from Renaissance to 1800, which is defined by the study of performance as pertaining to rhetorical aesthetics. The second phase is highlighted by Gustav Schilling's work *Musikalische Dynamik oder die Lehre vom Vortrag in der Musik* [94] in 1843. It is characterized by a switch from rhetorical aesthetics to the aesthetics of a musical work, a change that was initiated by Beethoven's creation of autonomous compositions, whose individual ideation would no longer be a subordinate instance of a general rhetorically expressive setup. The third phase is inaugurated by Hugo Riemann's work *Der Ausdruck in der Musik* in 1884 and can be characterized by the attempt to view performance as being an expression of analytical insights of the work *qua* meaningful text. Danuser, however, only includes analytical insights as rationales that are responsible for the performative shaping. He does not deal with gestures or emotional rationales, for example.

The first phase of rhetorical expressivity can be initiated with Nikolaus Listenius' *Musica...* [49] in 1537, dealing with the *science of right singing.* It appears parallel to Vasari's invention of the central perspective in painting in the last decade of the 15th century, which introduces the individual human position as opposed to the view of the world *sub specie aeternitatis*[1], the traditional Christian perspective of divine eternity. This new approach to performance stresses the individual position of the artist, and introduces human existence as a significant and valid point of view on the universe.

The breakthroughs of the human perspective of performance can be situated with Claudio Monteverdi's work in 1600, where performance is recognized as either expressing affects of the artist's musical person or inducing such in

[1] under the perspective of eternity

the audience. But the work has no essential individuality in this regard; it represents general categories of affect. These are transmitted in the framework of general rhetorics, more specifically according to the five classical rhetorical principles:

1. inventio (the arguments)
2. dispositio (the articulation)
3. elocutio (the communicative wording of thoughts)
4. memoria (memory for performance)
5. pronuntiatio/actio (the actual physical performance)

from which the last was the most important in rhetorically oriented performance.

Among the pedagogically oriented works on performance, the most prominents are: Johann Joachim Quantz: *Versuch einer Anweisung die Flöte traversière zu spielen* in 1752; Carl Philipp Emanuel Bach: *Versuch über die wahre Art das Clavier zu spielen* in 1753; and Leopold Mozart: *Versuch einer gründlichen Violinschule* in 1756, see also Figures 8.3.

Fig. 8.1: J. J. Quantz (1697-1773)

Fig. 8.2: C. Ph. E. Bach (1714-1788)

Fig. 8.3: L. Mozart (1719-1787)

These works as systematic, but not theoretically shaped. They apply the linguistic rhetorical tradition to the shaping of musical rendition. The progress is mainly in the written documentation of a culture of performance that used to be oral only, and which is to our days still strongly determined by oral education in the schools of music and conservatories. For example, Leopold Mozart describes and illustrates with images the different ways to hold the violin, either to lean it in a relaxed way on your shoulder, or else, in order to play more precisely, to fix it more tightly with your chin. And this also in order to create a specific posture for the social context in which performance is taking place. This all is also thought in the spirit of switching from the traditional understanding of instrumental performance as being an imitation of singing to an emancipated understanding of musical instruments. The human voice and language are being progressively relativized and complemented by specific and autonomous cultures of instrumental performance.

Fig. 8.4: D. G. Türk (1750-1813)

Fig. 8.5: A. L. Crelle (1780-1855)

Fig. 8.6: G. W. F. Hegel (1770-1831)

The explicitly theoretical discussion of performance appears with the works of Daniel Gottlob Türk: *Klavierschule* [108] in 1789; August Leopold Crelle: *Einiges über musicalischen Ausdruck und Vortrag* [14] in 1823; Georg Wilhelm Friedrich Hegel: *Aesthetik* [27] in 1835 (see Figure 8.6); Carl Czerny: *Klavierschule* [13] in 1840; and Gustav Schilling: *Musikalische Dynamik oder die Lehre vom Vortrage in der Musik* [94] in 1843 (see Figure 8.8).

Türk makes scientific remarks and explicitly addresses his book not only to students, but also to researchers. Crelle's book has a first clear semiotic structure. Performance is viewed as an exact expression of a work's content, as a whole and in parts. Here appears for the first time the idea of an individual and "autonomous" work character, which requires a specific treatment in order to achieve adequate expression of contents. This idea is germinating from Beethoven's creativity as being in opposition to general schemes, and as such requiring a thorough understanding and corresponding expression. Hegel goes one step further and distinguishes reproductive versus creative artistic production. He views performance as a recreation of the improvisational origins, and as such realizing the transformation of the composer's inside to the performative outside. Hegel's insights are astonishingly modern and remind us of Adorno's thoughts in [3]. Adorno views the ideal performance as a presentation of composed material as if it were improvised.

To these first theoretical writings one would add Carl Czerny's *Klavierschule* [13] in 1840 (Figure 8.7) and above all Gustav Schilling's *Musikalische Dynamik oder die Lehre vom Vortrage in der Musik* [94] in 1843 (Figure 8.8), which follows Hegel's systematization principles from general to particular. According to Schilling, performance has to comply with these three principles:

- objectivity: the reference to a completed composition when transforming it into sounding reality;

Fig. 8.7: Carl Czerny (1791-1857)

Fig. 8.8: Gustav Schilling (1805-1880)

- ideality: the role of the performer identifying and conveying the composition's ideas and contents;
- totality: capturing the totality of the ideas as expressed in the score's symbols.

Fig. 8.9: Hugo Riemann (1849-1919)

Fig. 8.10: Heinrich Schenker (1867-1935)

Although the theory is now an explicit topic when thinking of performance, it is still not clear which are the precise rationales of such musical shaping activity. The thematization of such rationales, and specificaly rationales from analytical inspection of musical compositions, was initiated by Hugo Riemann: *Der Ausdruck in der Musik* [87] in 1884 (Figure 8.11), then Heinrich Schenker: *Die Kunst des Vortrags* (unpublished) (Figure 8.12), and finally Theodor Wiesengrund Adorno: *Der getreue Korrepetitor* [2] in 1963, also in collaboration with Walter Benjamin.

Riemann specifies musical expressivity in performance as understood by the shaping of musical thoughts, the plastic relief of motives and themes, and the transparency of the entire construction of the work of art. He determines explicit rules of dynamics to shape analytical components such as melodies and harmonic modulations. Schenker argues for performance as an expression of analytical facts, introducing what Adorno will call analytical performance.

And finally Adorno discusses works of the second Viennese school, for example Webern's six bagatelles for string quartet. His thesis is that the performative shaping that does not descend to the specification of single notes of a composition is invalid as a performative rule. His rationales are derived from a strictly analytical point of view. Together with Benjamin, he coins the concept of a micrological procedure that penetrates the infinitely precise dimension of performative activity.

Fig. 8.11: Theodor W. Adorno (1905-1980)

Fig. 8.12: Walter Benjamin (1892-1940)

Summarizing his historical trajectory of philosophical performance theory, Danuser (loc. cit.) defines performance analysis (German: "Interpretationsanalyse") as being the translation of structural and formal insights (on the conceptual level of semiotic units) into performative shaping. Whereas rhetorical theory (German: "Vortragslehre," theory of artistic rendition) just applied general rules of rhetorics, such as *pronuntiatio*, to works of music, performance analysis fully recognizes the *specificity* of such works and expresses their contents in the shaping of performance, a methodology that became necessary with Beethoven's work.

8.1.2 The Empirical and Physical Performance Theory and Its Technology

8.1.2.1 First Empirical Approaches

For this thread of the history of performance theory, we refer to Reinhard Kopiez's discussion in [40]. He stresses the increasing role of the performer as an autonomous and core player in the creation and sounding realization of music, citing Franz Liszt's treatise *De la situation des artistes* (*ber die Stellung der Künstler*) [45] in 1835 (published in *Gazette musicale de Paris*), where Liszt equals composer, instrumental teacher, and performer as three variants of artistic personalities.

The historically traced empirical performance research described by Kopiez—and this is identical with the general history of empirical performance research—mainly deals with machines analyzing and/or synthesizing performance of keyboard music. This is due to the fact that keyboard technology is the only one traditionally accessible to precise measurements.

It is remarkable that empirical performance research goes back to 1745, when first construction plans for measuring key movements of the cembalo were made. In 1752, Johann Hohlfeld constructed a precursor of piano rolls, the *Fantasiermaschine* (Figure 8.13).

Fig. 8.13: Johann Hohlfeld (right) and the Klavier-Cembalo, a device built following Hohlfeld's *Fantasiermaschine*.

This machine was presented in 1753 to the Berlin Academy of the Sciences. It used pencils on paper, and ran 3.45 meters in 6.5 minutes. Carl Ph. E. Bach tested the device and approved it. The only sample was burnt in a fire at the academy. However we have a sample built by Jean-Joseph Merlin in London 1780, this one called *Klavier-Cembalo*.

In 1895, more than a century later, Alfred Binet and Jules Courtier built a machine with rubber air tubes measuring loudness and onset with one millisecond resolution (!), but not measuring pitch (Figure 8.14). The results from these measurements show that with increasing tempo the notes are played more legato and less regular in their temporal progression. This is visible from the sequence of five curves shown in Figure 8.14.

In 1932, Carl Seashore, the inventor of the Seashore Tests of Musical Ability (his name is translated from the Swedish surname Sjöstrand) invented the *Iowa Piano Camera* at the University of Iowa (Figure 8.15).

Seashore's results on agogics were groundbreaking. The machine had a resolution of 10 milliseconds and could measure relative note dynamics (in Figure 8.15: measured black block lengths from velocity of passing light J through window F on film D) in chords. These measurements proved agogical consistency of different renditions.

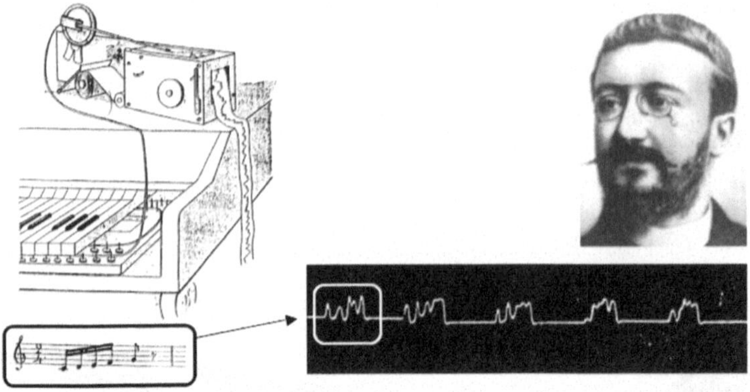

Fig. 8.14: Alfred Binet (right), his machine (left), and the loudness curve measuring several performances of the shown ascending motif.

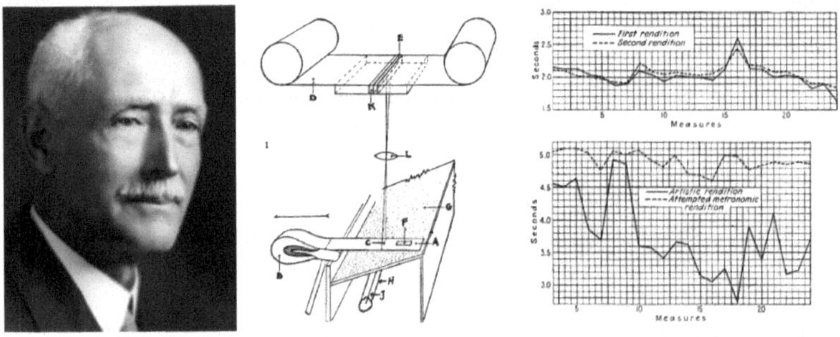

Fig. 8.15: Carl Seashore (left), his Iowa Piano Camera (middle), and the agogics curve (right) showing that agogics is not a random process.

After Seashore's death in 1949, empirical performance research paused until the Swedish research team of Erling Bløndal Bengtsson and Alf Gabrielsson started their SYVAR project in 1974 (completed by the studies of L. Henry Shaffer in 1980). Shaffer constructed an optoelectronic machine for a grand piano, feeding the hammer motions to a computer, a precursor of the modern disklavier. Finally, in 1983, the MIDI sound data format [?], [?] was introduced, and the recording of performance became general standard, although Christoph Wagner already had experimented with such devices with accuracy of 1 millisecond in 1974 [113].

Besides this arsenal of devices for the measurement of performances, we build upon the important tradition since 1883 of piano rolls for the analysis of historically important piano performances (see figure 8.16). There are piano rolls recorded by nearly all important pianists of the late nineteenth and early twentieth centuries. Their dynamics and timing is not very precise: They must

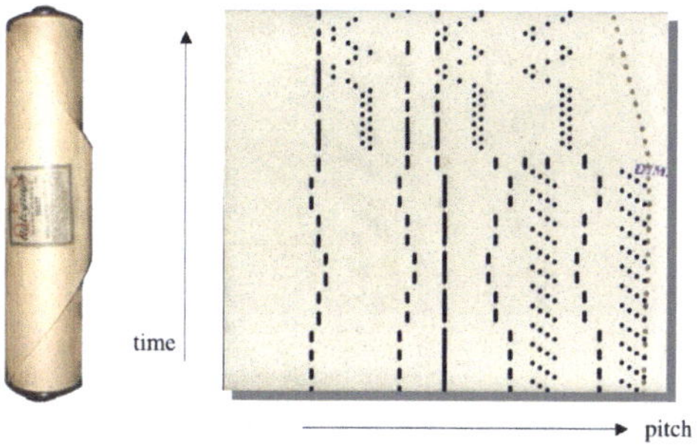

Fig. 8.16: Piano rolls are a historically important source of performance analysis between the last decade of the nineteenth and the first of the twentieth centuries.

be considered a simulation rather than an imitation of the performed music (see Figure 8.16).

Based upon this data, Rosina Seipp could prove in 1988 [95] that agogics are systematic rather than random (referring to the A-flat major polonaise by Frédéric Chopin). Around 1992, Hermann Gottschewski [22] had measured a great deal of piano rolls and found rules of agogics. Since 1994, there are MIDI readers for piano rolls by Horst Mohr [77], and also by Zotlan Jnosy and Janos Mcsai, but see [40]. Dynamics, however, is still difficult to measure precisely from these piano rolls.

A last method of empirical analysis of performances is the analysis of LPs and CDs. It has a long tradition and is coined "comparative interpretation/performance research". Here are some of the investigations conducted in this field: In 1926, Wilhelm Heinitz [40] compared timings from performances of Wagner's *Meistersinger*. In 1934, Adalbert Kalix [40] investigated ten recordings of Weber's *Freischue[U+009F]tz* overture, comparing performance with the score to find out about "musical romanticism." Different compositions, e.g. Beethoven's op.106 *Hammerklavier*, have been analyzed from LPs. Bruno Repp [84] analyzed recordings using FFT for precise onset calculations. In August 2008, Peter Neubäcker published *Melodyne*, a software [76] capable of transforming polyphonic acoustical material into symbolic score format (see Figure 8.17). The latter is a real revolution and will change empirical/comparative performance research dramatically.

The two threads of performance research were disjoint from each other in 1992, when my research group started its investigations about a comprehensive performance theory at the Computer Science Department of the University of Z"urich. At present, we are able to present a comprehensive theory, uniting the musicological/philosophical and the practical/empirical threads.

Let us conclude this overview with a very short remark about grammatical patterns of performance theory generated by machine-based learning from empirical performance data. We do not, in fact, believe that machine-delegated

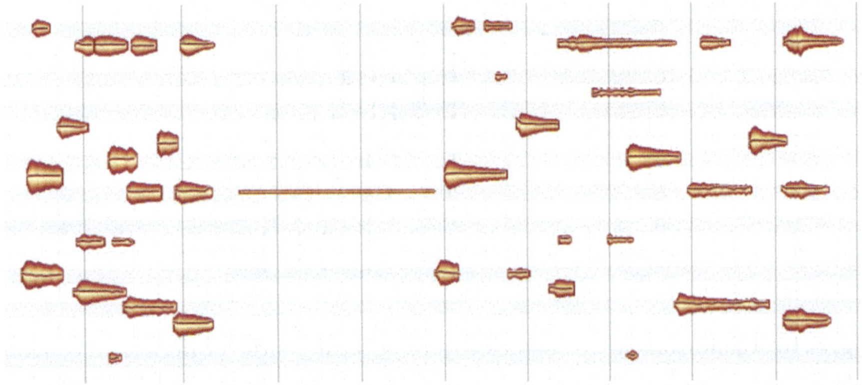

Fig. 8.17: The revolutionary software *Melodyne* is capable of transforming polyphonic acoustical material into symbolic score format.

statistical methods such as neural networks or proper machine learning algorithms for rule learning are of proper scientific value, since when machines learn, we do not. Of course, this is an ideological point of view, but we cannot follow methods that delegate decisions to structured ignorance; understanding cannot be delegated to engineered devices. For example, Gerhard Widmer's elaborate approach [116] starts with a relatively detailed structural analysis of the score, including motives, groupings, etc. It then correlates these structures to empirical performance data, such as dynamics or articulation, in order to apply machine learning algorithms for extrapolation to other scores. It however lacks the conclusions needed for the construction of a systematic performance grammar.

8.1.2.2 The Modern Physical Performance Theory

The modern physical performance theory is based upon the way we describe and analyze the quantitative relationship between symbolic parameters and their physical embodiment. For example, we will analyze the relationship between symbolic onsets and physical times of performed notes, a concept that is classically conceived as the performance's *tempo*.

Performance is defined as *a map $\wp$ associating the elements of a set C of notes of a composition with a corresponding set $\wp(C)$ of points in a physical space that represents those notes.*

The space of symbolic notes has six coordinates: onset o, pitch p, duration d, loudness l, glissando g, and crescendo c. The seventh coordinate of voice is left aside for our purposes. Each of these six coordinates will be a real number, element of the set $\mathbb{R}$ of reals. Therefore our composition will be a set C of points in $\mathbb{R}^6$.

Let us now describe the space of physical coordinates that correspond to C in performance. The physical onsets and durations are defined by seconds. The physical pitch is a bit more involved. It is related to the frequency of vibrating air. Frequency f means the number of times of a vibration that is performed per second. Human perception is given by the pitch $p(f)$ by a logarithmic formula as follows. To begin with, the logarithm $log(x)$ of a number x is given by the following idea: If we are given a number 10^l, its logarithm is the exponent l. For example, $1000 = 10^3$ is represented by 3, $log(1000) = 3$. Mathematicians can prove that for every positive real number x, there is a *logarithm* $log(x)$ such that $x = 10^{log(x)}$. And it tells us that $log(f.g) = log(f) + log(g)$. With this mathematical construction, *physical pitch* of frequency f is defined by the formula

$$pitch(f) = \frac{1200}{log(2)} log(f) + c$$

This formula represents in mathematical terms the way humans are perceiving air vibration. The pitch units are called *Cents*, *Ct*. Here is the meaning of this formula: If we apply it to the frequency $2f$ of one octave higher vibration, we get

$$pitch(2f) = 1200 log(2) log(2f) + c = pitch(f) + 1200$$

Therefore, $pitch(2f) - pitch(f) = 1200\ Ct$, the one octave pitch step is 1200 Cents, independently of the starting frequency. This is motivated by the octave being composed by 12 semitone steps, i.e., one semitone step is composed of 100 Ct steps.

The physical pitch unit Ct is also taken for the physical *glissando* coordinate. Physical loudness and *cresendo* are represented by a logarithmic formula similar to the pitch formula.

Therefore, the physical output $\wp(C)$ is also a set of points in $\mathbb{R}^6$. The performance $\wp$ then is a one-to-one correspondence

$$\begin{array}{ccc} C & \xrightarrow{\sim} & \wp(C) \\ \downarrow & & \downarrow \\ \mathbb{R}^6 & \xrightarrow{?} & \mathbb{R}^6 \end{array} \qquad (8.1)$$

between subsets of $\mathbb{R}^6$, the composition's set and it's performed physical set. The lower horizontal arrow, with a ?, stands for a virtually non-existent relation. This general image needs to be decomposed according to specifically musical perspectives, this is what we shall now develop.

9

Elements Of Performance

Summary. This chapter describes the conceptual components of performance, using modern mathematical structures.

– Σ –

9.1 Performance as a Gestural Map Between the Symbolic Score Space and the Physical Space of Sound Events

The first obvious approach would be to look for partial parameter spaces and the existence of corresponding factorization. Take a projection Pr onto a partial coordinate space and suppose that the bijection also takes place on this partial perspective:

$$\begin{array}{ccc} C & \xrightarrow{\sim} & \wp(C) \\ {\scriptstyle Pr}\downarrow & & \downarrow{\scriptstyle Pr} \\ Pr(C) & \xrightarrow{\sim} & Pr(\wp(C)) \end{array} \tag{9.1}$$

The musical idea is to look for projections onto musically specific subspaces. But let us first look at the mathematical basis for such a musical idea: performance vector fields. One typical example would be to look at the onset dimension, which means that the performance of symbolic onsets yields physical onsets, but no other dimension would be involved in this performance. For example, loudness or pitch are not involved in the performance of onsets.

G. Mazzola et al., *Basic Gestures for Music Theory, Performance, and Pedagogy*, Computational Music Science, https://doi.org/10.1007/978-3-032-10681-0_9

9.2 Performance Vector Fields

The structure of the performance bijection $\wp : C \sim \wp(C)$ can be understood when looking at its directional aspect. See Figure 9.1.

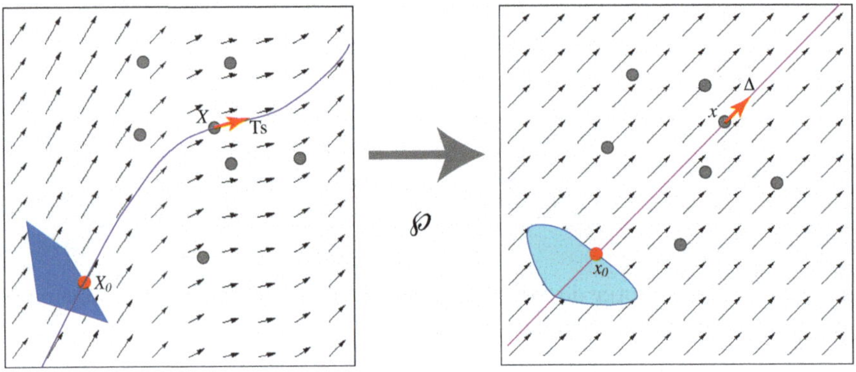

Fig. 9.1: The bijection between physical space and symbolic space induces a vector field correspondence with the diagonal field on the physical space. The blue sets are the sets of initial performance.

In this figure, we look at the vector field on the physical space, which is defined by diagonal vectors Δ of length 1 in every coordinate. The performance $x = \wp(X)$ of a point X is given by the image of the integral line through X, starting at an initial symbolic point X_0.

The vector field is also interpreted as a map between two gestures, see Figure 9.2

9.3 Special Situations: Tempo, Intonation/Tuning, and Articulation

9.3.1 Tempo

The most prominent such case is to project onto the onset spaces, symbolic and physical:

$$\begin{array}{ccc} C & \xrightarrow{\sim} & \wp(C) \\ {\scriptstyle Pr}\downarrow & & \downarrow{\scriptstyle Pr} \\ Onset(C) & \xrightarrow{\sim} & Onset(\wp(C)) \end{array} \tag{9.2}$$

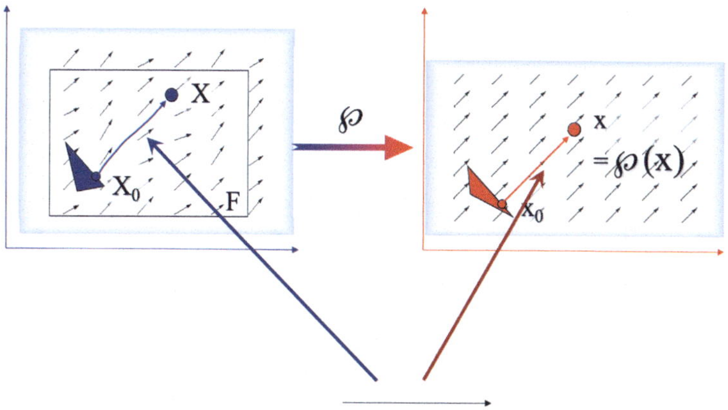

Fig. 9.2: The vector field formalism is also a map between two gestures.

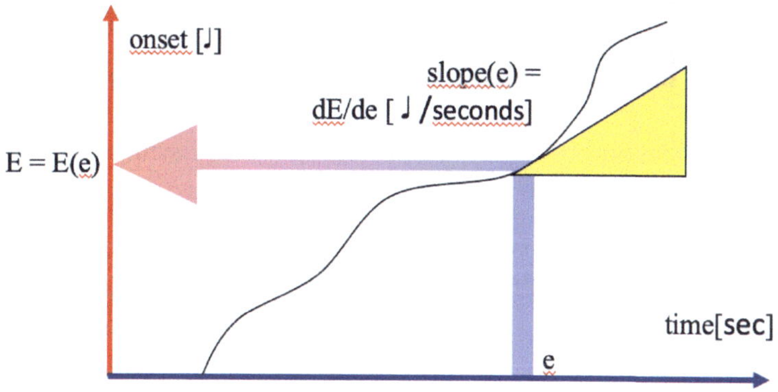

Fig. 9.3: The bijection between physical time and symbolic onsets is known as *tempo*.

The bijection on onsets can be represented in the direction from physics to symbols, see Figure 9.3. The slope of this bijection is called *tempo*, it is usually represented as a function of the onset value $E(e)$. For a car moving at a given speed, the slope would be "miles per hour". So you have tempo as an analogous reflection of "quarter notes per second".

9.3.2 Intonation/Tuning

Intonation or *tuning* is another one-dimensional performance case, this time between symbolic and physical pitch.

$$\begin{array}{ccc} C & \xrightarrow{\sim} & \wp(C) \\ {\scriptstyle Pr}\downarrow & & \downarrow{\scriptstyle Pr} \\ Pitch(C) & \xrightarrow{\sim} & Pitch(\wp(C)) \end{array} \tag{9.3}$$

which enables a functional representation as shown in Figure 9.4.

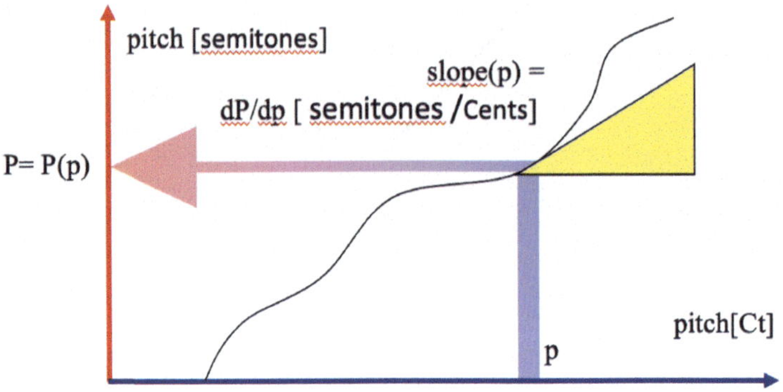

Fig. 9.4: The bijection between physical pitch and symbolic pitch.

The slope in Figure 9.4 is the *intonation*, a complete analogy to the tempo: intonation is the "tempo of pitch performance".

9.3.3 Loudness

Loudness is, like tempo, a one-dimensional projection

$$\begin{array}{ccc} C & \xrightarrow{\sim} & \wp(C) \\ {\scriptstyle Pr}\downarrow & & \downarrow{\scriptstyle Pr} \\ Loudness(C) & \xrightarrow{\sim} & Loudness(\wp(C)) \end{array} \tag{9.4}$$

which enables a functional representation as shown in Figure 9.5.

9.3.4 Articulation

Articulation refers to the shaping of duration, shortening it for *staccato* and lengthening it for *legto*. The difference to the afore mentioned parameters is that shaping duration also involves onset, it is a two-dimensional performance operator. We have this diagram now:

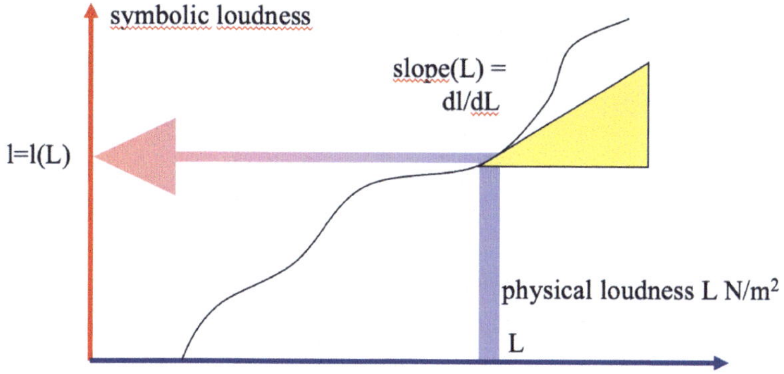

Fig. 9.5: The bijection between physical loudness (in terms of force per m^2) and symbolic loudness (in terms of p, mf, f, ff, fff, etc).

$$\begin{array}{ccc} C & \xrightarrow{\sim} & \wp(C) \\ {\scriptstyle Pr}\downarrow & & \downarrow{\scriptstyle Pr} \\ Onset + Duration(C) & \xrightarrow{\sim} & Onset + Duration(\wp(C)) \end{array} \quad (9.5)$$

The $\wp$ map sends the pair (E, D) to $(e(E), d(E, D) = e(E + D) - e(E))$. For example, the symbolic onset and duration could be $(3/4, 1/4)$, while its performance would be $(performance - of - 3/4), performance - of - 4/4)$. The corresponding slope is

$$Z(E, D) = \partial T(E, D) = (T(E), 2T(E + D) - T(E))$$

Figure 9.6 shows the performance vector field of the above slope.

We may consider a more general set of articulation fields defined by

$$\partial_\lambda(E, D) = (T(E), (1 + \lambda^{-1})T(E + D) - T(E)$$

where $\lambda = 1$ is the previous situation. Figure 9.7 shows three values $1, 0.3, 3$ of λ

9.4 Generating Meaningful Performance

Performance can be expressive in three mean directions: emotional, gestural, and analytical.

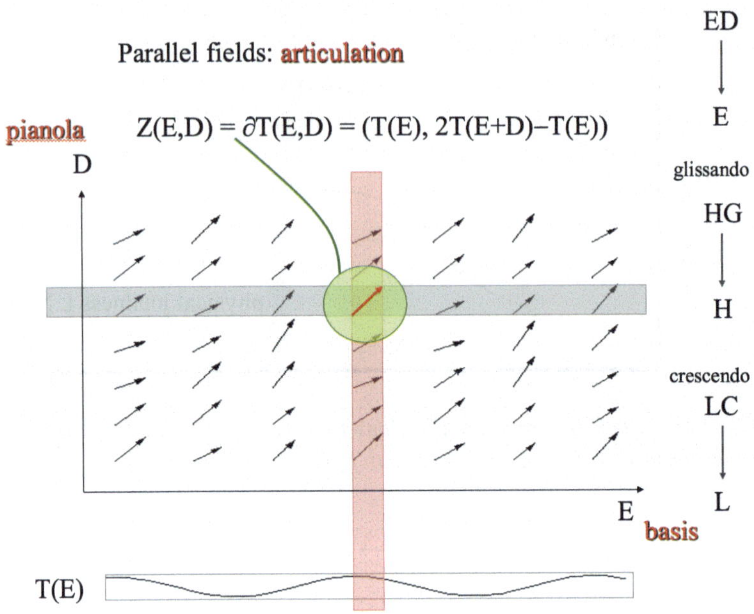

Fig. 9.6: The articulation slope $\partial T(E, D)$. We show the two dimensions: the basis E and the pianola dimension D; but the same structure shows up for crescendo C and glissando G.

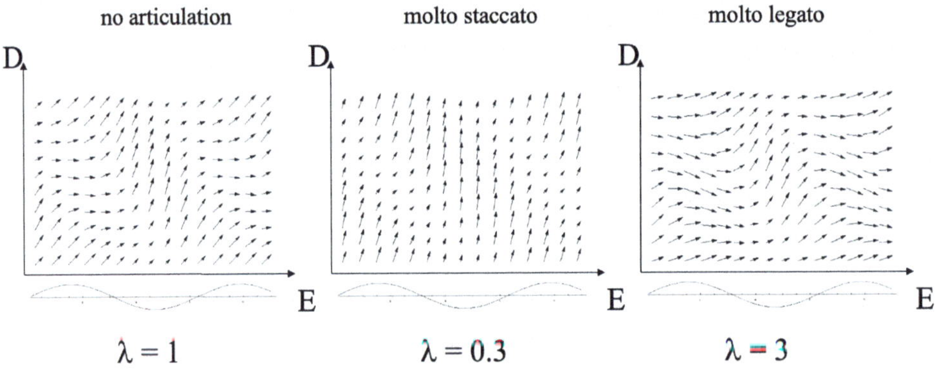

Fig. 9.7: The articulation slope for three values $1, 0.3, 3$ of λ.

9.4.1 Emotional Performative Gestures

In ths section, we refer to the work of Renate Wieland [109], a piano pedagogue and student of the famous performance theorist Theodor Wiesengrund Adorno. In collaboration with her colleague Jürgen Uhde, she has described the emotional content of piano gestures in a geometric language:

> Musical gestures are perceived in the free conducting movement, in the playing movement and sublimated in the spiritual mimesis of pure imagination. Whatever the level, such experiments are always within space. Originally, affects were actions, related to an exterior object, along the process of interiorization they were detached from their object, but they are still determined by the coordinates of space.
> (...)
> Language reminds us everywhere of the connection of affect and movement and of the way gestures behave in space.
> (...)
> The touch of sound is the target of the comprising gesture; the touch is so-to-speak the gesture within the gesture at large, it equally relates to the coordinates of space.

Wieland, when referring to "affect", reminds us of the etymology of "emotion", namely derived from the Latin "ex-movere", to move from inside out. This is her theory of emotional expressivity of musical gestures. And in the last part of our quotation, she even explains what we call a hypergesture in 2.1.4: a gesture of gestures.

9.4.2 Gestures as Performative Expression

The second meaningful expressivity of performance is realized by musical gestures as ultimate contents, without any further semantics. This situation is prominent in Beethoven's compositions. He builds score symbols that mean to be gestures without any further content. For example, in Beethoven's Sonata op. 57, "Appassionata", in the middle of the first page we recognize a strong downward gesture, that ends with a halting fermata, see Figure 9.8

Fig. 9.8: The strong downward gesture at the beginning of Beethoven's op. 57.

Glenn Gould, who is an excellent Bach performer, did not appreciate Beethoven, and we recognize this attitude in Gould's performance of Beethoven's op. 57, where Gould avoids every possible gesture and produces a simply stupid performance.

9.4.3 Analytical Expressivity in Performance

This third type of gestural expressivity in performance has been investigated in a creative collaboration with statistician Jan Beran [63, Part XI]. We could use the precise measurments of tempo in Schumann's famous composition " Träumerei", composition 7 in Schumann's "Kinderszenen". These measurements were provided to us by Bruno Repp [63, ref.887], including performances of 28 famous pianists. We provided the analytical functions for rhythmical, melodic, and harmonic structures associated with Schumann's composition. Using advanced statistical methods, we could prove that the Repp functions are significantly correlated to the analytical functions, i.e., these famous pianists shape the tempo in "Träumerei" in a statistically significant relationship with the analytical structures.

This correlation could be confirmed for some other prominent compositions, see [63, Part XI] for details.

9.5 String Quartet Theory

The string quartet, with first and second violin, viola, and violoncello, is a classical instrumental display. We want to discuss why these four instruments are essential as a performative group. There are two components of this display: the music theory being realized and the sound variety that these instruments can propose.

The theoretical basis of classical string quartets is a double: the environment of tonal modulations together with the counterpoint structure at the end of the 18th century.

The sound variety of a string quartet is defined by the multiplicity of sounds that are enabled by these violin instruments.

The result of this configuration is that for the violin instrumentation, we need four or more instruments to enable a presentation of those harmonic and contrapuntal structures, see [71, Chapter 19]. It is in fact remarkable that several prominent composers (Haydn for example) had difficulties composing with more than four violins.

10

Creating Musical Time

Summary. Musical time is a phenomenon within performance, which is different from physical time and can be conceived as a gestural utterance.

– Σ –

10.1 Adorno and Kramer

In his famous book [1, p.247], Theodor Wiesengrund Adorno describes performance as a gestural utterance:

> Correspondingly the task of the interpreter would be to consider the notes until they are transformed into original manuscripts under the insistent eye of the observer; however not as images of the author's emotion—they are also such, but only accidentally—but as the seismographic curves, which the body has left to the music in its gestural vibrations.

And he adds:

> All making music is a recherche du temps perdu (on the search of lost time) [82].

In other words, the gestures are a mechanism in search of a lost time. We are reconstructing musical time by means of our performing gestures.

Jonathan Kramer in his book [41] has displayed a theory of musical time, which could help making Adorno's claims more precise. He writes:

> In the following chapters I postulate many types of musical time. To do so becomes possible once I accept the notion that music *creates* time. (...) Time itself can (be made) move, or refuse to move, in more than one 'direction': not an objective time out there, beyond ourselves, but the very personal time created within us as we listen deeply to music. (...)

G. Mazzola et al., *Basic Gestures for Music Theory, Performance, and Pedagogy*, Computational Music Science, https://doi.org/10.1007/978-3-032-10681-0_10

> The difference between ordinary lived time and musical time is, according to Clifton, the difference "between the time a piece takes and the time which a piece presents or evokes." (...) All the species of time discussed in this book are experienced and understood simultaneously with ordinary time.
>
> Kramer then displays a fourfold variety of time concepts:
>
> - gestural time: species of multiply-directed musical time in which the conventional meaning of gestures (beginnings, endings, structural upbeats, etc.), rather than the literal successions of events, determine the logic of continuity.
> - moment time: temporal continuum of a moment-form composition
> - multiply-directed (linear) time: temporal continuum in which progression is seemingly in several directions at once
> - vertical time: temporal continuum of the unchanging, in which there are no separate events and in which everything seems part of an eternal present.

The most dramatic time here is "vertical time", where the linearity of physical time is abolished in favor of a spatial simultaneity. The Adorno and Kramer approaches suggest that musical time should be a gestural process.

10.2 Our Construction of a Gestural Musical Time

From our analysis of musical collaboration in jazz improvisation, we learn that it results in a collective "distributed identity". The gestural structure of this identity results from the following concepts. We are given a system of gestures (G_i) on the given composition. For some pairs G_i, G_j of these gestures, there is a gestural map $f_{i,j} : G_i \to G_j$. We shall now define, what is a gestural map. But the important mathematical result is, that there exists a gesture $\lim(G_i)$, the *limit* of (G_i), together with a gesture map $p_i : \lim(G_i) \to G_i$ for each i. The limit is the "smallest" gesture with this property: Every other such gesture H admits a gesture map from the limit to H. This limit is what we need to define the musical time structure of the system.

What is a gestural map? Given two gestures $g : S \to X^{\to}, h : T \to Y^{\to}$, a map $f : g \to h$ is a map of directed graphs $f_1 : S \to T$ plus a continuous function $f_2 : X \to Y$ such that $f_2^{\to} \circ g = h \circ f_1$. Figure 10.1 shows the map on the skeleta.

The following figures show three examples of such limits

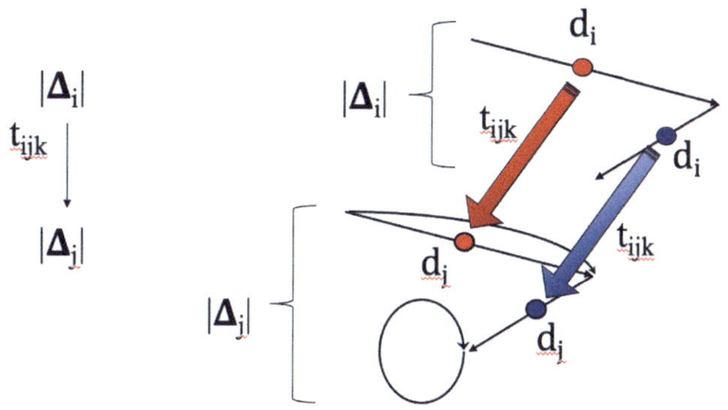

Fig. 10.1: The map on the skeleta Δ_i, Δ_j.

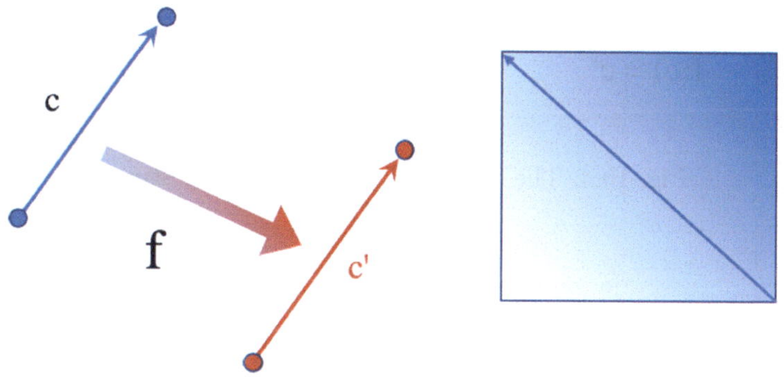

Fig. 10.2: The limit of two simple gestures.

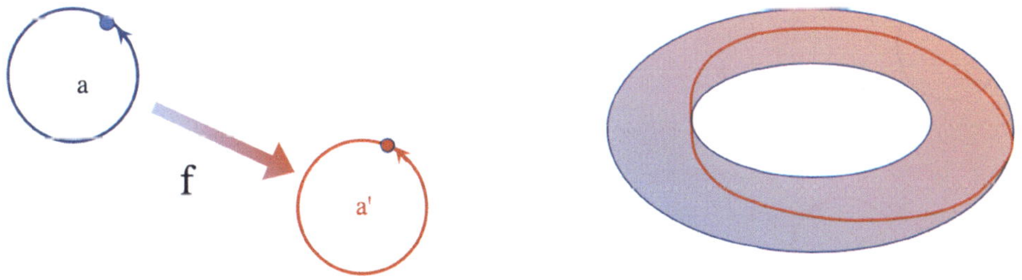

Fig. 10.3: The limit of two circular gestures. The product of two circles is a donut, and the map associates pairs of points of these two circles in the donut.

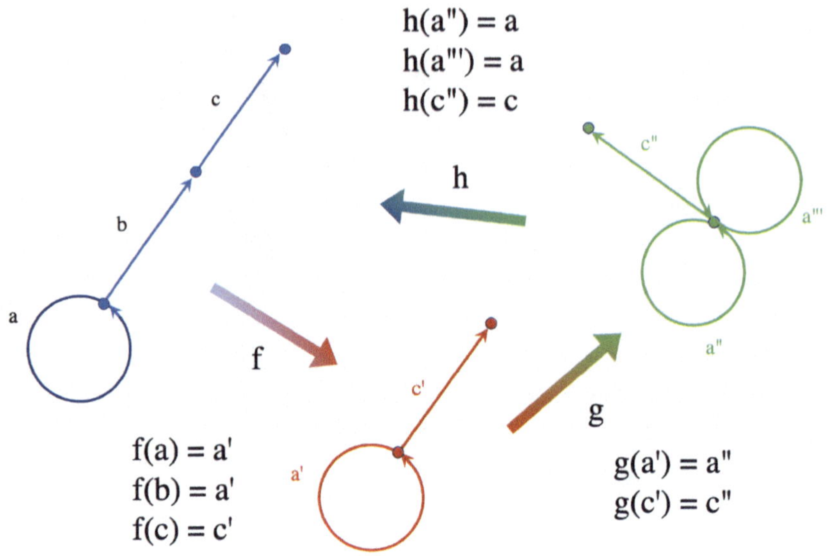

Fig. 10.4: The system of more involved gestures.

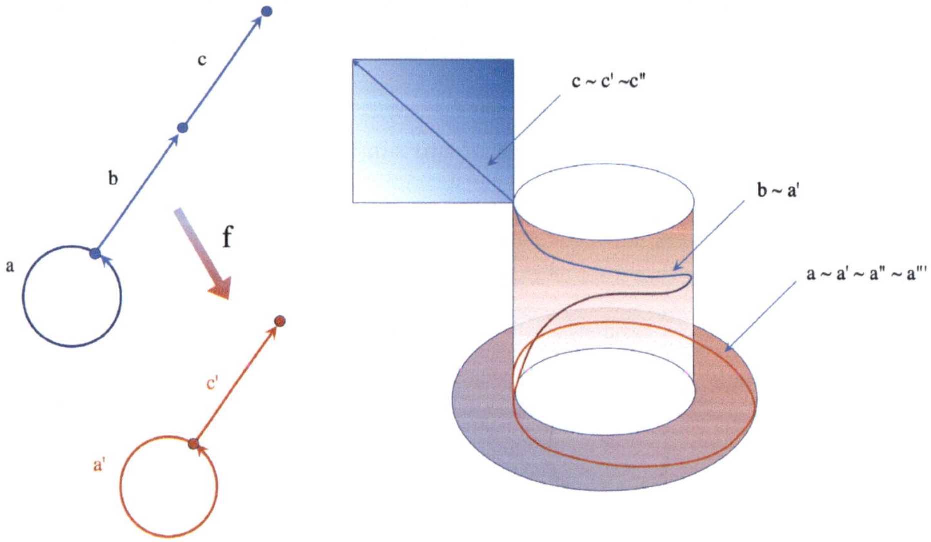

Fig. 10.5: The limit of a system of more involved gestures.

Part IV

Contents

11

Contents In Music

Summary. This chapter discusses the methods and tools to generate contents in music.

– Σ –

11.1 What Contents do We Have?

We have investigated musical contents of the following types: Emotional, gestural, and analytical expressivity, as well as the creation of musical time.

These types are not mutually exclusive. Musical time, for example, will be instantiated together with any of the other contents. There are three types of performance constructors:

- *symbolic* constructors, which change the initial symbolic information. For example: separating right from left hand in a piano score.
- *performance* constructors, which influence the transition from symbolic to physical space. For example: the tempo operator, which generates a new tempo.
- *physical* constructors, which change the given physical output. For example: change the loudness of the given performance.

These constructors may be defined by abstract logic or by an improvisational approach. *But the overall impetus is the human desire to build fictitious realities as alternatives to "real life" or as experimental proposals for new directions.*

We say that we *play* music, because it is the action of a game in our fantasy, a serious game in performance.

G. Mazzola et al., *Basic Gestures for Music Theory, Performance, and Pedagogy*, Computational Music Science, https://doi.org/10.1007/978-3-032-10681-0_11

Part V

Pedagogy

12

Pedagogy

Summary. This part and chapter discusses pedagogical principles associated with our gestural methodology.

– Σ –

12.1 Introduction

Music pedagogy is about the theory and practice of musical education. In view of our gestural methodology, it is natural to add a pedagogical exposé of its educational realization. Music pedagogy is not a new field, there are a number of recognized approaches, such as eurhythmics by Émile Jaques-Dalcroze [15], Zoltàn Kodály's method [39], Carl Orff's Schulwerk [79], or Shinichi Suzuki's method [103].

All these methods stress the performative perspective of music and have been successful approaches to early education. They all necessarily stress instrumental embodiment, and Delacroze specifically focuses on movement in time (rhythm). Nevertheless, none of these successful approaches is conceived around gestural processuality. Our pedagogical display will expose this last educational perspective.

Besides the processual endeavor, making music is also a dynamic that is spread over two different spatial realities: the abstract space of musical symbols (notes, score architectures, etc.) plus the physical space of sound events in time. This double spatiality is connected by a sophisticated transformational vector: the action of playing those symbols in performance. This action is the target of musical gestures, i.e., basically, playing, making music involves symbols, sounds, and the performative transformation of symbols into sounds.

Music pedagogy has to conceive and instruct this logic in an evolutionary unfolding. In particular, it would be pernicious to teach only a part of this triple reality (gestures, symbols, sounds), music would be misrepresented and conceptually destroyed.

Here are the initial gestures:

G. Mazzola et al., *Basic Gestures for Music Theory, Performance, and Pedagogy*, Computational Music Science, https://doi.org/10.1007/978-3-032-10681-0_12

12.1.1 Performance of One Note

In Figure 12.1, we show the initial pedagogical gesture that realizes the performance of a note on a piano, the note being with pitch 67 = g, onset 0, duration 1/4, loudness *mf* (not shown in the figure), glissando and crescendo both 0, voice = piano.

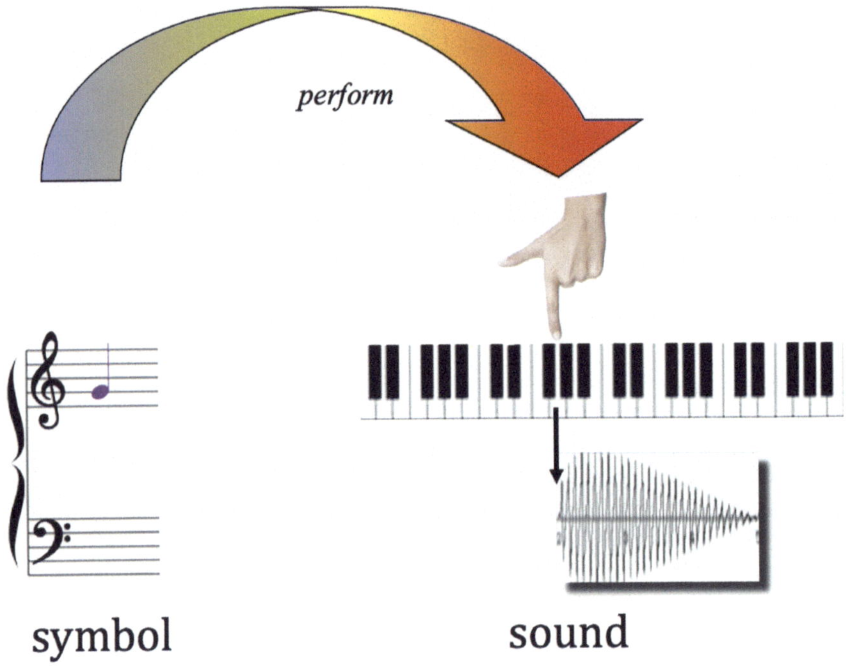

Fig. 12.1: The performance gesture of a single note.

It is useful to perform also on other instruments, such as the violin, where glissando and crescendo can be played by a movement on the string. This initial exercise is important to teach the differences and commonalities between symbols and sounds. The physical parameters of the sounds may be different, and should be critically investigated. However, we should also teach the two spaces as relatively independent realities, which collaborate in performance.

12.2 Developing Music Theory and Performance in Parallel

The performance and representation of one note may be enriched by the same processuality for the developing music theory. One may go on with the conception of intervals, chords, tonalities, modulations and contrapuntal structures, everything in a gestural language. — But one may also, as the desire for performative richness of symbolic data increases, investigate successively more complex performative perspectives and approaches. This might be a function of the students' preferences. We have displayed music theory and performance in the previous parts of this book. Students (especially children) might benefit from a variety of different combined approaches to increase understanding. For example, understanding the concept of an interval might be eased when playing the interval instead of its abstract, conceptual description.

12.3 Creativity Pedagogy

The theoretical and performative activity should also be understood under the light of creativity: These techniques are not interesting if distinguished from creative instances. This pedagogical aspect must be realized, when the students build symbolic structures with their performative realizations.

The realization of such projects would be enhanced if one could instantiate music software for score notation, e.g., *Sibelius*, *Finale*, *GarageBand*, or *Logic*. This technology must be considered with regard to the students' options.

Behind all these more or less technical aspects, we should not forget about the deeper question of "what is music". This is not about how we may use music to express some perspective of our life, such as sex or power. It is about the substance of music *as such*. It is the same as it is wrong to give a list of human organs and bones, when asked what is a human. This challenge could be a strong background motivation for delving into music as a young student.

In an overall perspective, music is that projection in performance of a system of symbols. Music Theory is the architecture of those symbols. Its substance is a free construction of rationales for the performative projection.

References

1. Adorno Th W: Fragment über Musik und Sprache. Stuttgart, Jahresring 1956
2. Adorno Th W: Der getreue Korrepetitor (1963). Gesammelte Schriften, Bd. 15, Suhrkamp, Frankfurt am Main 1976
3. Adorno Th W: Zu einer Theorie der musikalischen Reproduktion. Suhrkamp, Frankfurt/M 1946
4. Augustinus: Thesaurus Musicarum Latinarum. School of Music, Indiana University, Bloomington
5. Bentley A: Fixed or Movable Do? Journal Of Research In Music Education 1959
6. Lang S: Algebra. Springer 2002
7. Agustín-Aquino O A, J Junod, G Mazzola: Computational Counterpoint Worlds. Springer Series Computational Music Science, Heidelberg 2015
8. Astley R: Rick Astley—never gonna give you up (official music video). `https://youtu.be/ dQw4w9WgXcQ?si=n6CL9FBeSsvVmO27`. Performance of Never Gonna Give You Up by Rick Astley, 2009.
9. Bevilacqua F et al.:F. On Designing, Composing and Performing Networked Collective Interactions. Organised Sound, Cambridge University Press (CUP), 2021, 26 (3), pp.333-339.
10. Boulez P: Musikdenken heute I,II; Darmstädter Beiträge V, VI. Schott, Mainz 1963, 1985
11. Boulez P: structures, premier livre. UE, London 1953
12. Boulez P: structures, deuxième livre. UE, London 1967
13. Czerny C: Klavierschule. 1840
14. Crelle A L: Einiges [U+009F]ber musicalischen Audruck und Vortrag. 1823
15.
16. Danuser H et al. (eds.): Neues Handbuch der Musikwissenschaft, Bd. 11: Musikalische Interpretation. Laaber, Laaber 1992
17. Du Y: Chinese Traditional Music Theory Course. Shanghai Music Publishing House 2004
18. Eimert H: Grundlagen der musikalischen Reihentechnik. Universal Edition, Wien 1964
19. Funkhouser Ch: being matter ignited...an interview with Cecil Taylor. Hambone 1994
20. Fux J J: Gradus ad Parnassum (1725). Dt. und kommentiert von L. Mitzler, Leipzig 1742; English edition: The Study of Counterpoint. Translated and edited by A Mann. Norton & Company, New York, London 1971

G. Mazzola et al., *Basic Gestures for Music Theory, Performance, and Pedagogy*, Computational Music Science, https://doi.org/10.1007/978-3-032-10681-0

21. Fügedi J: Dance Notation and Computers, Yearbook for Traditional Music, Volume 23, 1991
https://www.thegreatcoursesdaily.com/kurt-godel-wanted-to-revise-our-concept-of-time
22. Gottschewski H: Tempohierarchien. Musiktheorie, Heft 2, 1993
23. Graeser, W: Bachs "Kunst der Fuge". In: Bach-Jahrbuch 1924
24. Hanslick E: Vom Musikalisch-Schönen. Breitkopf und Härtel (1854), Wiesbaden 1980
25. Hatol 2006
26. Hecquet S and A Prokhoris: Fabriques de la danse. PUF, Paris 2007
27. Hegel G W F: Vorlesungen [U+009F]ber die Aesthetik (1835). Werke in 20 B[U+008A]nden mit Registerband: 13: Vorlesungen Â[U+009F]ber die Aesthetik I: BD 13, suhrkamp, Berlin 1986
28. Hjelmslev L: Prolégomènes á une théorie du langage. Minuit, Paris 1966
29. Ho L and Han K: On Chinese Scales and National Modes. University of Texas Press 1982
30. Huang H: Theory and practice in the traditional Chinese music: Observations and analysis. University of North Texas 1994
31. Hudson R: Stolen time : the history of tempo rubato. Oxford: Clarendon Press (1995), ISBN 0198161697.
32. Jakobson R: Linguistics and Poetics. In: Seboek, TA (ed.): Style in Language. Wiley, New York 1960
33. Jackendoff R and Lerdahl F: A Generative Theory of Tonal Music. MIT Press, Cambridge MA, 1983
34. Jie L et al: Yinyue Jianshang. Anhui Literature and Art Publishing House 2018
35. Johnson I and Henrich A K: Knot Theory. Aurora 2017
36. Kaiser J: Beethovens 32 Klaviersonaten und ihre Interpreten. Fischer, Frankfurt/Main 1979
37. Kant I: Kritik der reinen Vernunft. Meiner, Hamburg 1956
38. Kendon A: Gesture: Visible Action as Utterance. Cambridge University Press 2004
39.
40. Kopiez R: Aspekte der Performanceforschung. In: de la Motte H: Handbuch der Musikpsychologie. Laaber, Laaber 1996
41. Kramer J D: The Time of Music. Schirmer, New York 1988
42. Lee B and Mazzola G: Interdisciplinary Approaches to Musical Time. 2nd International Conference MUSICA ANALITICA, Coimbra, Portugal, Oct. 2023
43. Lewin D: Generalized Musical Intervals and Transformations (1987). Cambridge University Press 1987
44. Ligeti G: Pierre Boulez: Entscheidung und Automatik in der Structure Ia. Die Reihe IV, UE, Wien 1958
45.
46. Jin J: Chinese Music. Cambridge University Press 2011
47. Kelley J L: General Topology. Van Nostrand 1955
48. Kraus R: The Cultural Revolution: A Very Short Introduction. Oxford University Press 2012
49. Listenius N: Musica... 1537
50. Kassia: Rachmaninoff - Piano Concerto No.2, I. Moderato. https://youtu.be/r-SZ_e5GWMc?si=2sVgMWL89JyAI_TH. Performance of Piano Concerto No. 2 by Kassia, 2020.

51. Kuttner F: Prince Chu Tsai-Yü's Life and Work: A Re-Evaluation of His Contribution to Equal Temperament Theory. University of Illinois Press 1975
52. Mac Lane S: Categories for the Working Mathematician. Springer 1998
53. Mac Lane S and Moerdjik I: Sheaves in Geometry and Logic. Springer 1992
54. Maris B: Making Music At The Piano. Oxford University Press 2000
55. Murowchick R: China: Ancient Culture, Modern Land. University of Oklahoma Press 1994
56. Maisels C: The Near East: Archaeology In The 'Cradle of Civilization'. Routledge 200
57. Marek C: Lehre des Klavierspiels. Atlantis 1977
58. Mazzola G: Gruppen und Kategorien in der Musik. Heldermann, Berlin 1985
59. Mazzola G: *Synthesis.* SToA 1001.90, Zürich 1990
60. Mazzola G: `www.rubato.org`
61. Mazzola G et al.: The Topos of Music—Geometric Logic of Concepts, Theory, and Performance. Birkhäuser, Basel et al. 2002
62. Mazzola G et al.: The Topos of Music I: Theory. (First volume of second edition of [61]) Springer, Heidelberg 2018
63. Mazzola G et al.: The Topos of Music II: Performance. (Second volume of second edition of [61]) Springer, Heidelberg 2018
64. Mazzola G et al.: The Topos of Music III: Gestures. (Third volume of second edition of [61]) Springer, Heidelberg 2018
65. Mazzola G et al.: The Topos of Music IV: Roots. (Fourth volume of second edition of [61]) Springer, Heidelberg 2018
66. Mazzola G, G Milmeister, J Weissmann: Comprehensive Mathematics for Computer Scientists, Vols. I, II. Springer, Heidelberg et al. 2004
67. Mazzola G et. al.: Making Musical Time. Springer 2021
68. Mazzola G and M Andreatta: Diagrams, Gestures, and Formulas in Music. Journal of Mathematics and Music 2007, Vol. 1, no. 1, 2007
69. Mazzola G and P B Cherlin: Flow, Gesture, and Spaces in Free Jazz—Towards a Theory of Collaboration. Springer Series Computational Music Science, Heidelberg 2009
70. Mazzola G: Categorical Gestures, the Diamond Conjecture, Lewin's Question, and the Hammerklavier Sonata. Journal of Mathematics and Music Vol. 3, no. 1, 2009
71. Mazzola G: Musical Performance—A Comprehensive Approach: Theory, Analytical Tools, and Case Studies. Springer Series Computational Music Science, Heidelberg, December 2010
72. Mazzola G, J Park, F Thalmann: Musical Creativity. Springer, Heidelberg 2011
73. Mazzola G, M Mannone, Y Pang: Cool Math for Hot Music. Springer, Heidelberg 2016
74. Nan N and Guan X: Common and distinct quantitative characteristics of Chinese and Western music in terms of modes, scales, degrees and melody variations. Journal of New Music Research 2023
75. Noll Th: http://www.cs.tu-berlin.de/ noll/ChordDictionary.sea.hqx, TU Berlin 1996
76. `http://en.wikipedia.org/wiki/Celemony#Melodyne`
77. `http://ns1.foxtail.com/Tech/mohrReader.html`
78. Nattiez J-J: Fondements d'une Sémiologie de la Musique. Edition 10/18 Paris 1975

79. Perlman I: John williams: Schindler's list theme - itzhak perlman. https://youtu.be/cLgJQ8Zj3AA?si=oq5loL5k0YX-lI1x. Performance of Schindler's List Theme by Itzhak Perlman, 2019.
80. Pfeifer R and Ch Scheier. Understanding Intelligence. MIT Press 1999
81. Proust M: A la recherche du temps perdu. e-artnow 2014
82. Rahaim M: Musicking Bodies. Wesleyan University Press, Middletown CT 2012
83. Repp B: The Sound of Two Hands Clapping: An Exploratory Study. J,. Acoust. Soc. Am. 81 (4), April 1987
84. Riemann B (1867): Über die Hypothesen, welche der Geometrie zugrunde liegen (Habilitationsreferat 1854). Gott. Abh. No.13, 1867
85. Riemann H: System der musikalischen Rhythmik und Metrik. Breitkopf und Härtel, Leipzig 1903
86. Riemann H: Musikalische Dynamik und Agogik. Leipzig 1884
87. Ries I: dance movement.
88. Pang Y: Rural And Urban Explorations Of Transnational Performance: A Dialogue Between Traditional Southwestern Chinese And Contemporary North American Music And Dance. University of Minnesota 2018
89. Rachmaninoff S: Sergei Rachmaninoff plays his Piano Concerto No. 2. https://youtu.be/x8l37utZxMQ?si=xV-FL56bddd9FZBL. Performance of Piano Concerto No. 2 by Sergei Rachmaninoff, 1929.
90. Ruwet N: Langage, Musique, Poesie. Seuil, Paris 1972
91. Sachs K-J: Der Contrapunctus im 14. und 15. Jahrhundert. AMW, Franz Steiner, Wiesbaden 1974
92. Schönberg A: Harmonielehre (1911). Universal Edition, Wien 1966
93. Schilling G: Musikalische Dynamik oder die Lehre vom Vortrag in der Musik. 1843
94. Seipp R: Analyse von Interpretationen mit dem Welte-System: Aufgezeigt an der Agogik der Polonaise A-Dur von Chopin. Das Mechanische Musikinstrument vol. 12; issue 44, 1988
95. Singh
96. Seekhunlio W: Yangqin: The origin and development in China: Origin and development in China. International Journal of Curriculum and Instruction 2023
97. Shu Z and Shu H: Xinbian Shuzhao Erhu Course. Chengdu Times Press 2004
98. Singh
99. Song F: "The Sun Shines on Taxkorgan" by Chen Gang, Song Fei Erhu Solo. https://www.youtube.com/watch?v=3eVTrgbR2Gw (2005)
100. Straub H: Beiträge zur modultheoretischen Klassifikation musikalischer Motive. Diplomarbeit ETH-Zürich, Zürich 1989
101. Sun H: Performance of The Sun Shines on Tashkurgan by Sun Huang. https://www.bilibili.com/video/BV1pe41137jf/ (2022)
102.
103. Tsao P: Tradition and Change in the Performance of Chinese Music. Harwood Academic Publishers 1998
104. Tsui Y: The Modern Chinese Folk Orchestra: A Brief History. Musical Performance 1998
105. Tweed ?, 2009
106. Tymoczko D: A Geometry of Music. Oxford University Press, 2011
107. Turk D G: Klavierschule. 1789
108. Uhde J and Wieland R: Denken und Spielen. Bärenreiter, Kassel et al. 1988

109. Uhde J and R Wieland: Forschendes Üben. Bärenreiter 2002
110. Valéry P: Cahiers I-IV (1894-1914). Celeyrette-Pietri N and Robinson-Valéry J (eds.), Gallimard, Paris 1987
111. Varèse E: Erinnerungen und Gedanken. In: Darmstädter Beiträge III. Schott, Mainz 1960
112. Wagner Ch and H von Karajan: Experimentelle Untersuchungen Über das Tempo. Oesterreichische Musikzeitschrift 29, 1974, S. 589-604
113. Wang G: Wang Guotong Erhu Mingqu Yanshi. Shanghai Music Publishing House 2020
114. Watts, V: Benesh Movement Notation and Labanotation: From Inception to Establishment (1919–1977). Dance Chronicle, 38(3), 275–304. https://doi.org/10.1080/01472526.2015.1085227
115. Widmer G: Modeling the Rational Basis of Musical Expression. Computer Music Journal Vol. 18, 1994
116. Wendorff R: Zeit und Kultur. Westdeutscher Verlag, Opladen 1980
117. Wikipedia: gesture https://en.wikipedia.org/wiki/Gesture
118. Yust J: Stylistic information in pitch-class distributions. Journal of New Music Research 2019
119. Zhao H: Zhongguo Erhu Qudian. Hunan Literature and Arts Publishing House 2020
120. Zhu Z: Convergence: Unveiling the Yanyue Modal System of the Tang Dynasty. Rice University 2021
121. Zoon J: Density 21.5. Edgard Varèse: The Complete Works. CD. Decca London Polygram 289 460 208–2,
https://www.youtube.com/watch?v=s6-TBeDLlnA 1998

The manufacturer's authorised representative in the EU is Springer
Nature Customer Service Centre GmbH, Europaplatz 3, 69115 Heidelberg,
Germany. If you have any concerns regarding our products, please
contact ProductSafety@springernature.com

Printed and bound by CPI Group (UK) Ltd, Croydon, CR0 4YY
05/10/2026
02228059-0001